T0344402

Alternative Medicines for Diabetes Management

Apart from diet and exercise, the strategic use of different classes of prescribed or non-prescribed xenobiotic compounds for the restoration of euglycaemic levels in the body is well known. The ongoing rivalry between the recommended usage of allopathic medicines versus Ayurvedic remedies has encouraged many researchers to focus their studies on thoroughly isolating and characterizing extracts from different parts of plants and then evaluating their relative activities via *in vitro*, *in vivo*, and, in some cases, clinical studies.

Alternative Medicines for Diabetes Management: Advances in Pharmacognosy and Medicinal Chemistry provides a holistic view of all oral therapies for diabetes mellitus that are available to the public by removing the silos and stigmas associated with both allopathic and Ayurvedic medicines.

Additional features include

- Highlights the potential role of dietary and medicinal plant materials in the prevention, treatment, and control of diabetes and its complications.
- Educates readers on the benefits and shortcomings of the various current and potential oral therapies for diabetes mellitus.
- Allows the quick identification and retrieval of material by researchers learning the efficacy, associated dosage, and toxicity of each of the classes of compounds.
- Presents the history, nomenclature, mechanisms of action, and shortcomings of each of the various subclasses of allopathic therapeutants for diabetes mellitus and then introduces Ayurvedic medicines.
- Chapter 4 discusses various metallopharmaceuticals and provides a holistic view of all available and potential therapies for the disease.

Medicinal Plants and Natural Products for Human Health

Series Editor:

Christophe Wiart

We are at a point in history where the global population is increasingly interested in medicinal plants, natural products and herbalism. This interest has accelerated with COVID-19 and long COVID symptoms. Rediscovery of the connection between plants and health is responsible for a new generation of botanical therapeutics that include plant-derived pharmaceuticals, multicomponent botanical drugs, dietary supplements, functional foods and plant-produced recombinant proteins. Many of these products can complement conventional pharmaceuticals in the treatment, prevention and diagnosis of diseases. This series is a critical reference for anyone involved in the discovery of biopharmaceuticals for the improvement of human health.

Alternative Medicines for Diabetes Management: Advances in Pharmacognosy and Medicinal Chemistry

Edited by: Varma H. Rambaran, Nalini K. Singh

Alternative Medicines for Diabetes Management

Advances in Pharmacognosy and Medicinal Chemistry

Varma H. Rambaran and Nalini K. Singh

University of Trinidad and Tobago, Tamana Campus,
Trinidad and Tobago

CRC Press is an imprint of the
Taylor & Francis Group, an **informa** business

First edition published 2022
by CRC Press
6000 Broken Sound Parkway NW, Suite 300, Boca Raton, FL 33487-2742

and by CRC Press
4 Park Square, Milton Park, Abingdon, Oxon, OX14 4RN

CRC Press is an imprint of Taylor & Francis Group, LLC

Library of Congress Cataloging-in-Publication Data

Names: Rambaran, Varma H., author. | Singh, Nalini K., author.
Title: Alternative medicines for diabetes management : advances in pharmacognosy and medicinal chemistry / Varma H. Rambaran, Nalini K. Singh, University of Trinidad and Tobago, Tamana Campus, Trinidad and Tobago.
Description: First edition. | Boca Raton, FL : CRC Press, 2023. | Includes bibliographical references and index.
Identifiers: LCCN 2022017667 (print) | LCCN 2022017668 (ebook) | ISBN 9781032344898 (hbk) | ISBN 9781032344973 (pbk) | ISBN 9781003322429 (ebk)
Subjects: LCSH: Diabetes--Alternative treatment. | Diabetes--Ayurvedic treatment. | Materia medica, Vegetable--Therapeutic use. | Medicinal plants--Therapeutic use. | Pharmacognosy.
Classification: LCC RC661.A47 R36 2023 (print) | LCC RC661.A47 (ebook) | DDC 616.4/62--dc23/eng/20220716
LC record available at https://lccn.loc.gov/2022017667
LC ebook record available at https://lccn.loc.gov/2022017668

ISBN: 9781032344898 (hbk)
ISBN: 9781032344973 (pbk)
ISBN: 9781003322429 (ebk)

DOI: 10.1201/9781003322429

Typeset in Times
by Deanta Global Publishing Services, Chennai, India

For my mentors:

Dr Richard A. Fairman and Professor Alvin A. Holder

Thank you for your continued guidance and support

Varma

Contents

Preface

Apart from diet and exercise, the strategic use of different classes of prescribed and non-prescribed xenobiotic compounds for the restoration of euglycaemic levels in the body is well known. The ongoing rivalry between the recommended usage of allopathic medicines versus herbal remedies has encouraged many researchers to focus their studies on thoroughly isolating and characterizing extracts from different parts of plants, and then evaluating their relative activities via *in vitro*, *in vivo*, and, in some cases, clinical studies. To further support this drive, the respective rich histories of many countries with regard to both colonization and immigration have enabled a wider selection of botanical sources to become more accessible to their inhabitants, and it is observed that these plants are already integrated into their regular diets and lifestyles.

Stepping aside from the aforementioned controversy between fact and folklore, the emergence of a revolutionary class of therapeutics that belongs to a family of compounds known as coordination complexes has also been reported. Fondly dubbed by their developers as insulin mimetics or insulin enhancers, these metallo-drugs have struggled for acceptance by the medical fraternity due to the divisive claims behind them. However, perseverance driven by passion in this area has continued to yield promising data, and this has stoked hope for their further development.

Acknowledgements

The authors wish to thank the University of Trinidad and Tobago for all the assistance given throughout the development of this book.

We also wish to thank our families and friends for their undying encouragement and support throughout our endeavours.

Finally, we would like to give special thanks to Mrs Ashmini Motilal, Miss Pritivi Narine, Miss Kajol Jagessar, Mr Dheeresh Ramcharan, and Mr Narendra Maharaj for their invaluable contributions to the completion of this book.

Authors

Dr Varma H. Rambaran is currently employed as an Associate Professor at the University of Trinidad and Tobago (UTT), Office of the Vice President of Research and Student Affairs (VP-RASA). He graduated from the University of the West Indies in 1999, with a Bachelor of Science in Chemistry, and in 2005, obtained a Doctor of Philosophy degree in Inorganic Chemistry. During his years as a post-graduate student, he developed a keen interest in the behaviour of coordination complexes in biological systems. However, it was during his post-doctoral fellowship at the University of Southern Mississippi that the ameliorative effects of vanadium-based complexes in diabetes therapy caught his attention. Through the support of the International Centre for Genetic Engineering and Biotechnology (ICGEB), he was able to successfully explore and develop his idea of using a family of novel vanadyl complexes (PDOV and PYTOV) as insulin-enhancing agents. The findings from his studies were awarded a patent by the US Patent and Trademark Office in March 2021.

While working on his project, Dr Rambaran noticed a partiality to certain medicines that were being prescribed to diabetic patients. The seemingly undying controversy over the superiority of allopathic medicines, versus their ethno-pharmaceutical counterparts has resulted in the polarization of many groups, due to their lack of awareness and understanding of the different forms of therapies. To further complicate matters, the majority of reference books on the market commonly lack the supporting scientific data of compound identity, mechanism of action, and safety in use. Unfortunately, this has led to a bias against the use of herbal medicines by insinuating that the claims being made are more witchcraft than science.

Inspired by this, Dr Rambaran saw a need to furnish a comprehensive book of this nature, which would impartially and holistically educate its readers on the oral therapeutic options that are available to them.

Nalini Kathleen Singh is currently employed at UTT as a Research Assistant, under the Office of VP-RASA. Ms Singh's research interests fall under the umbrella of non-communicable diseases (NCDs), focusing mainly on the epidemiology and etiology of hypothyroidism. Her extended interest in natural medicines used in NCD-therapies has warranted her invaluable contributions in this book.

Ms. Singh holds a Bachelor of Science in Chemical and Process Engineering, which she obtained from the University of the West Indies (UWI, St Augustine Campus) in 2013. She also completed an Occupational Safety and Health Administration (OSHA) general industry training course at the School of Business and Computer Science (SBSC) in 2014, and participates in short courses on a regular basis to improve her professional skills. Since 2016, she has been an active member of the Association of Professional Engineers of Trinidad and Tobago (APETT) in both the electrical and chemical divisions, with her current grade being Associate Member. Ms Singh has also received awards in leadership and good citizenship owing to her involvement in volunteer programmes since 2002. Due to her positive attitude, her desire to learn, and her passion for scientific research, Ms Nalini Singh continues to take steps to continuously improve, both professionally and personally.

1 Etiology

1.1 DIABETES MELLITUS (DM)

Diabetes mellitus is ranked as one of the top four non-communicable diseases in the world. According to its most recent release, the International Diabetes Federation (IDF) has estimated a world population of 537 million people (aged 20–79 years) who are living with the disease, which, if left unaddressed, will rise to a staggering 784 million by 2045! Within the North American and Caribbean region alone there are 51 million adults, representing a regional prevalence of 9.5% (Figure 1.1); this prevalence had an associated health expenditure of approximately US$415 billion and accounted for 42.9% of total global diabetes-related health expenditure in 2021 (IDF Diabetes Atlas 2021: 10th Edition 2021).

DM's diagnosis as a metabolic disease arises from either the pancreas's inability to produce sufficient quantities of insulin or the desensitization of the human insulin receptor glycoprotein (INS-R) (due to the overproduction of the hormone). Its etiology encompasses a range of factors that include genetics, viral infections, autoimmunity, and obesity (Ramasamy and Schmidt 2014). Stemming from these are its clinical manifestations, which include hyperglycaemia, glycosuria, insulin resistance, fat and carbohydrate metabolism abnormalities, and chronic complications resulting from macro- and micro-vascular pathology. DM can be categorized under two major classes: type-1 and type-2. However, the diagnosis of a patient with DM and the classification under which type they suffer from, can vary according to the individual. As such, it has been observed in recent times that endocrinologists have further expanded the subclasses according to their contrasting occurrences and symptoms. An example of this is a patient who was initially diagnosed with gestational diabetes (GDM) and continued to be hyperglycaemic even after delivery and, as a consequence, was later diagnosed with type-2 diabetes mellitus (T2DM) (IDF Diabetes Atlas 2021: 10th Edition 2021). To best understand this and other occurrences, a brief discussion of the various classifications of the disease follows.

1.1.1 Type-1 Diabetes

Type-1 diabetes mellitus (T1DM) or "immune-mediated diabetes" was previously classified as "insulin-dependent diabetes" or "juvenile-onset diabetes." It occurs as a result of the cellular-mediated autoimmune destruction of the pancreatic β-cells, which may be either genetically or environmentally influenced. However, there has been some suspicion of its occurrence due to toxic and some dietary factors (Sesti 2006, Nelson and Cox 2008). The rate of destruction of the β-cells has been observed to vary from slow (occurring mainly in adults) to rapid (being highly peculiar to infants and children). The condition can develop at any age, although it occurs most frequently in children and young adults.

DOI: 10.1201/9781003322429-1

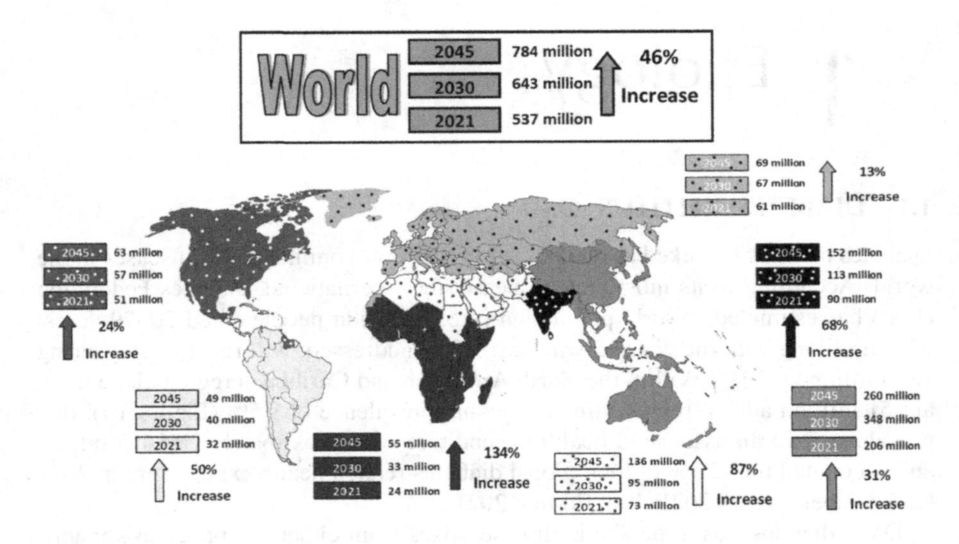

FIGURE 1.1 International Diabetes Federation Diabetes Atlas (2021).

The incidence of T1DM is increasing worldwide, but there is considerable variation by country with some regions of the world having much higher incidences than others. The reasons for this are unclear, but the rapid increase over time is suspected to be due to non-genetic factors, such as environmental and probably lifestyle-related changes. Apart from this, proposals of an observed inverse relationship between the decreasing incidence of infections in Western countries and the increasing incidence of both autoimmune and allergic diseases have been put forward to justify the spike (Okada, Kuhn and Feillet 2010).

Therapeutic treatments for people with T1DM include daily insulin injections, which when complemented with regular blood glucose (BG) monitoring, education, and support, allow the patient to live a relatively normal life.

1.1.2 Type-2 Diabetes

Previously classified as "non-insulin-dependent diabetes" or "adult-onset diabetes," T2DM typically encompasses patients who either display a resistance to insulin or have problems with secreting the hormone. During the state of insulin resistance, the hormone is ineffective and in due course prompts an increase in insulin production. Consequently, the β-cells become overworked and eventually cannot keep up with the demand for the hormone.

Specific etiologic factors for T2DM are unknown; however, β-cell destruction (as in the case of T1DM) does not occur. This type predominantly occurs in adults and children who are obese, which by itself is a contributing factor to insulin resistance. Interestingly, individuals who may not be considered obese by the traditional criteria of weight, tend to exhibit an increased percentage of body fat distribution, mostly to the abdominal area (American Diabetes Association 2013).

Globally, the prevalence of T2DM is high and rising across all regions. This rise is driven by population ageing, economic development, and increasing urbanization leading to more sedentary lifestyles and a greater consumption of unhealthy foods linked to obesity (Sesti 2006). The cornerstone of T2DM management is the promotion of a lifestyle that includes a healthy diet, regular physical activity, smoking cessation, and maintenance of a healthy body weight. As a contribution to improving the management of T2DM, in 2017 the IDF issued the IDF Clinical Practice Recommendations for Managing T2DM in Primary Care (Aschner 2017).

If attempts at lifestyle changes are not sufficient to control blood glucose levels, oral medication is usually initiated with biguanides (metformin) as the first line of therapy. Should this treatment prove to be inadequate or inefficient, a range of combination-therapy options, such as sulfonylureas, meglitinides, and dipeptidyl peptidase-4 inhibitors are then considered for prescribed use. Finally, as a last resort, if oral medications fail to control the hyperglycaemic levels, insulin injections may be deemed necessary (IDF Diabetes Atlas 2021: 10th Edition 2021).

1.1.3 Hyperglycaemia in Pregnancy

Hyperglycaemia in pregnancy (HIP) can be classified as either "gestational diabetes mellitus" (GDM) or "diabetes in pregnancy" (DIP). It can occur anytime during pregnancy, including the first trimester; however, its occurrence is usually detected during the third trimester. DIP classically applies to pregnant women who have previously known diabetes or have hyperglycaemia that was first diagnosed during pregnancy and meets the World Health Organization's (WHO) criteria of diabetes in the non-pregnant state.

Due to hormone production by the placenta, GDM also arises in women with an insufficient insulin-secretory capacity to overcome insulin resistance. The risk factors for GDM include older age (over 35 years), overweight and obesity, previous cases of GDM, excessive weight gain during pregnancy, a family history of diabetes, polycystic ovary syndrome, habitual smoking, and a history of stillbirth or giving birth to an infant with a congenital abnormality (IDF Diabetes Atlas 2021: 10th Edition 2021).

1.1.4 Other Types of Diabetes

According to a recently published WHO report on the classification of diabetes mellitus, there are a number of "other specific types" of diabetes, including "monogenic diabetes" and what was once termed "secondary diabetes."

As the name implies, monogenic diabetes results from a single gene rather than the contributions of multiple genes and environmental factors as seen in T1DM and T2DM. With an occurrence of only 1.5%–2%, these cases are noted to be in the minority; however, they include various manifestations that range from neonatal diabetes mellitus (sometimes called "monogenic diabetes of infancy"), maturity onset diabetes of the young (MODY), and rare diabetes-associated syndromic diseases (IDF Diabetes Atlas 2021: 10th Edition 2021).

1.2 INSULIN TRANSDUCTION PATHWAY

The insulin transduction pathway refers to a self-regulating biochemical process in which the hormone (insulin) influences the uptake of glucose into the fat and muscle cells or conversely reduces the synthesis of glucose in the liver.

The release of insulin from the pancreas is a consequence of spiked levels of glucose in the blood. Identified as the main contributor to normoglycaemic regulation, the released insulin promotes the uptake of glucose by binding to its membrane-embedded receptor and triggering a cascade of intracellular processes that promote the usage and/or storage of glucose in the cell.

To better appreciate the action of insulin and its potential substitutes, a closer look at its involvement in the insulin transduction pathway is necessary.

Spanning the cellular membrane is the insulin substrate receptor (INS-R). This glycoprotein is composed of two extracellular α-subunits, which are linked to their complementary β-subunits via disulphide bonds. The β-subunit itself comprises three sections: the extracellular, transmembrane, and intracellular domains; the latter section contains three main autophosphorylation sites: the juxtamembrane domain (JM), the catalytic domain (TK), and the COOH-terminal phosphorylation domain (CT) (Figure 1.2) (Sesti 2006).

As illustrated in Figure 1.3, after insulin enters the bloodstream it binds to the α-subunits of the INS-R and forms a receptor–ligand complex (Step 1). This, in turn, triggers the intracellular domain of the receptor's β-subunit to undergo phosphorylation by adenosine triphosphate (ATP) (Step 2). Now decorated with the phosphate groups, the INS-R becomes a suitable site for the transfer of these phosphate groups

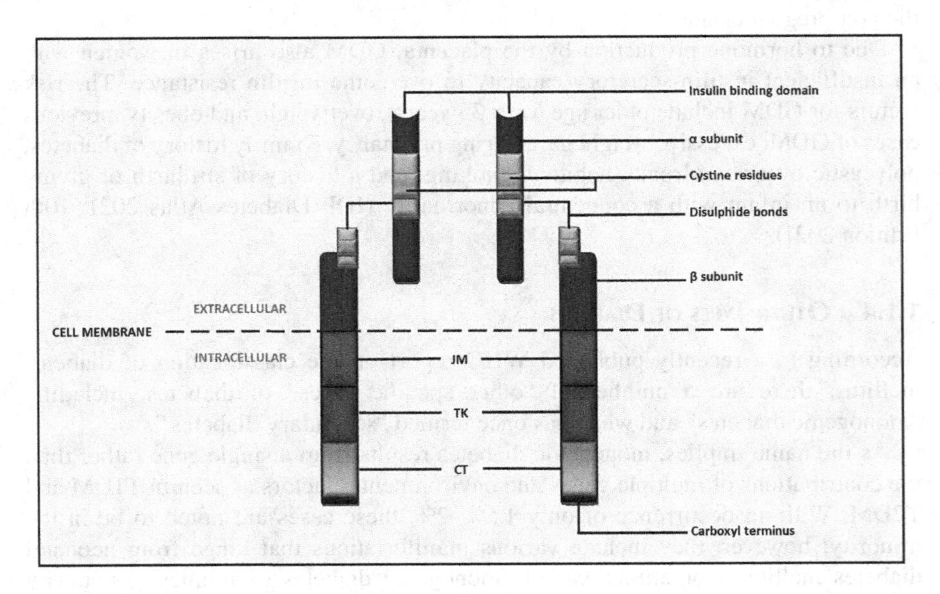

FIGURE 1.2 Diagrammatic representation of the human insulin receptor glycoprotein (INS-R).

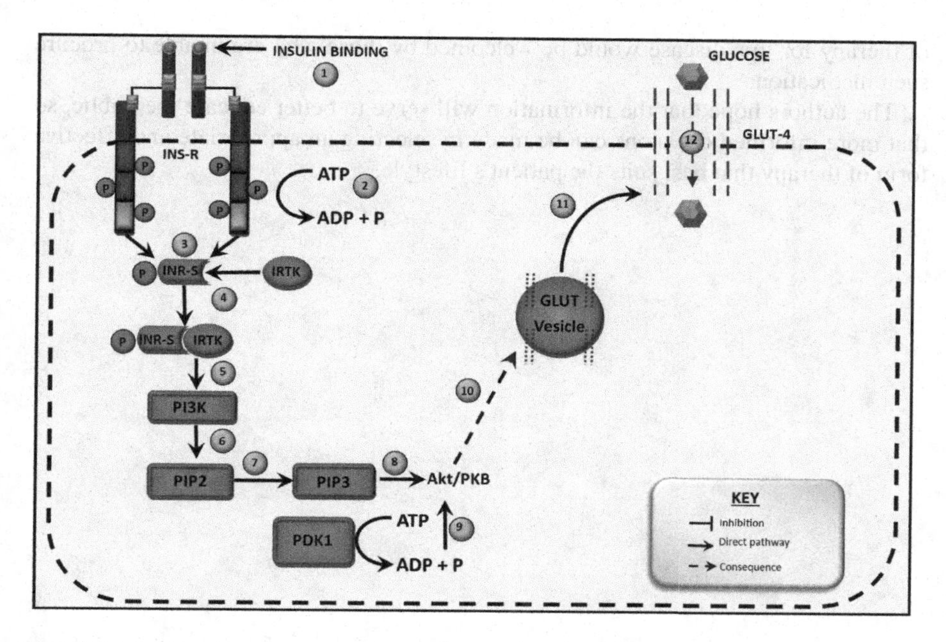

FIGURE 1.3 Schematic diagram of the insulin transduction pathway.

to other target proteins, such as the insulin receptor substrate (INR-S) (Step 3). The phosphorylated INR-S is now a target for insulin receptor tyrosine kinase (IRTK), which upon binding (Step 4) activates the enzyme phosphatidylinositol 3-kinase (PI3K) (Step 5). This domino effect continues as the activated PI3K binds to the membrane lipid, phosphatidylinositol (4,5)-bisphosphate (PIP2) (Step 6) and converts it to phosphatidylinositol (3,4,5)-triphosphate (PIP3) (Step 7). Next in the sequence is protein kinase-B (PKB/Akt), which binds to PIP3 via pleckstrin homology (Step 8). PKB is then phosphorylated on a threonine residue by phosphoinositide-dependent kinase-1 (PDK1) (Step 9), which in turn signals the translocation of the intracellular vesicle glucose transporter type-4 (GLUT-4) to the cell membrane (Step 10). The GLUT-4 finally fuses with the cell membrane (Step 11) and creates a portal through which glucose can enter (Step 12).

1.3 OUR GOAL

There is a large pool of reliable and compelling publications that are based on the justification for using either an entire class of oral therapeutics or individual members of that particular class. However, there does not exist any form of compiled reference that impartially reports the benefits and shortcomings of all these forms of therapy.

We were also concerned about the uncontrolled rise in cases of DM throughout the world and the lack of availability of many prescription drugs, due to economic factors. We believe that knowledge (backed by science) of alternative forms of therapy for this disease would be welcomed by those who are unable to procure such medication.

The authors hope that the information will serve to better educate the public, so that more informed decisions can be made in selecting an appropriate and effective form of therapy that best suits the patient's lifestyle.

2 Allopathic Medicines

Let's see...Apart from Headaches, Fever, Shortness of Breath, Temporary Blindness and in the rare case, Death...

I'd say that the drugs I gave to you are pretty much safe.

Despite the many confirmed and unconfirmed side effects, allopathic drugs are still peddled as the safest form of therapy.

2.1 INTRODUCTION

2.1.1 Overview and Significance

Regarding allopathic medicines, drugs are strictly defined as chemical substances that are used to prevent or heal diseases in humans, animals, and plants, and are grouped according to the manner in which they are utilized. Of particular interest to the medicinal chemist is the structure of the compound as well as its relative pharmacological mechanism of action. However, other categories such as the nature of the illness must also be considered.

In the case of diabetes mellitus (DM), there is a need for a multi-targeted strategy for its efficient management. For example, postprandial hyperglycaemia (at the digestive level) is controlled with α-glucosidase inhibitors such as acarbose, miglitol, and voglibose, which are used to inhibit the degradation of carbohydrates. Alternatively, to enhance glucose uptake by peripheral cells, biguanides such as metformin are

DOI: 10.1201/9781003322429-2

seen to be the therapeutant of choice, while insulinotropic sulfonylureas such as glibenclamide serve as secretagogues for pancreatic cells (Nimesh, Tomar and Dhiman 2019).

The structure–activity relationship (SAR) approach to drug discovery is based on the observation that compounds with a structural resemblance to a known pharmacologically active drug are often themselves similarly active. This activity may be either comparable to that of the lead drug, with differences in potency and toxicity, or may be completely different. Appreciatively, a study of the SARs between a lead compound and its analogues may be used to determine the respective moieties of the parent structure that are responsible for both its beneficial biological activity and also its unwanted side effects. This information may be used to develop a new drug that has improved activity and fewer unwanted side effects.

2.1.2 History

The term "allopathy" was coined by the inventor of "homeopathy," the late Samuel Hahnemann, and was used to describe the somewhat barbaric procedures that were being performed by his colleagues at that time. Being an advocate of "observation and experiment" versus "tradition and opinion," Hahnemann condemned allopathic practices, and used them as examples to point out how physicians, with conventional training, were employing ineffective therapeutic techniques. He believed that if a high dose of a therapeutic substance caused symptoms similar to those of a disease in a normal person, then it may alleviate the symptoms of such a disease when given in high dilution. As such, there can be an established relationship between the dilution of the substance and its potency.

Despite Hahnemann's rantings, allopathic techniques continued to be practiced. Although critics of modern medicine in the 20th century implied that only immunization was inconsistent with the singularity of allopathy and conventional medicine, the practice was observed to increase in prevalence amongst main stream physicians, possibly due to the interest in complementary and alternative remedies. (Gundling 1998).

The continued interest in allopathic therapies led to the introduction of synthetic compounds to be used as therapeutic drugs, in the late 19th century. This later spurred the exploration of new remedies, which were less toxic than the naturally sourced substances that were in use. The development was based on the modelling of novel structures (analogues) that bore a physical resemblance to known pharmacologically active compounds (leads), and then experimentally assessing their relative activities and toxicities to gauge their superiority in action (Thomas 2007).

2.1.3 Downfalls

Despite the many benefits of allopathic medicines, their deficiencies must also be deliberated. Firstly, we should consider holistically the associated toxicity with this form of treatment. All drugs, both prescription and non-prescription, are considered poisonous once taken in excess. For example, an overdose of the over-the-counter drug paracetamol can cause coma and death. Additionally, drug resistance must also be considered, as it affects the ability of the body to respond to certain medicines.

Another point is that therapies developed along the principles of allopathy are often limited in efficacy, and are often too costly, especially for developing countries like India (Nimesh, Tomar and Dhiman 2019). Finally, allopathic medicines are known to carry the risk of having adverse events, and are generally prescribed based on risk versus benefit for a particular disease and patient (Paudyal et al. 2019) (Table 2.1).

2.2 BIGUANIDES

Biguanides, as exemplified by one of its more popular members metformin (Figure 2.1), refer to the family of insulin-enhancing drugs with the general formula: $HN(C(NH)NH_2)_2$. In the early 1920s, guanidine was found to be the active component of the plant *Galega officinalis* Linn. The isolate was reported to exhibit remarkable hypoglycaemic activity and was later used as the lead drug in the synthesis of several antidiabetic compounds (Rena, Hardie and Pearson 2017, Bailey, Biguanides and NIDDM 1992).

These compounds commonly improve insulin sensitivity by decreasing hepatic glucose production, decreasing intestinal absorption of glucose, and increasing peripheral glucose uptake and utilization. The multiple pathways through which metformin affects liver metabolism have been described in detail by Rena et al. (Rena, Hardie and Pearson 2017). However, due to the highly technical nature of the overlapping pathways, more simplified sites of action of the drug are presented in Figure 2.2. In the first instance (Step 1), the compound is catalytically taken into the hepatocytes by the organic cation transporter-1 (OCT1). There, the drug accumulates in the cells and further in the mitochondria, due to the similarity in membrane potentials. Within the mitochondria, the drug inhibits Complex I, which initially prevents mitochondrial adenosine triphosphate (ATP) production (Step 2). This inhibition increases cytoplasmic [adenosine diphosphate (ADP) : ATP] and [adenosine monophosphate (AMP) : ATP] ratios, which in turn activates AMP-activated protein kinase (AMPK) activity. The increased AMP : ATP ratios also result in multiple indirect inhibitory effects, which include the suppression of fructose-1,6-bisphosphatase (FBPase) action (Step 3); a consequence of which is the acute inhibition of gluconeogenesis. Another inhibitory effect, arising from the altered AMP : ATP

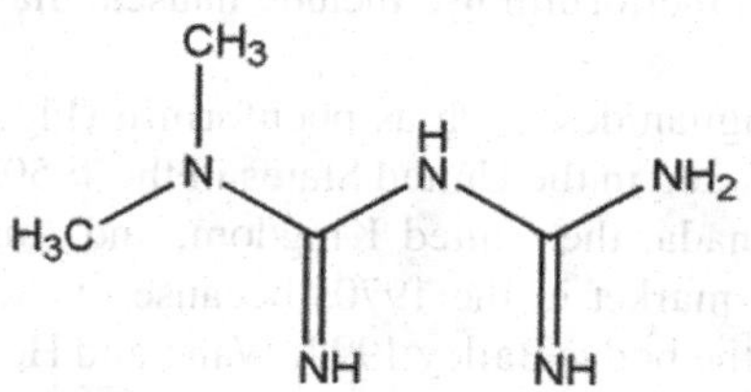

FIGURE 2.1 Structure of metformin.

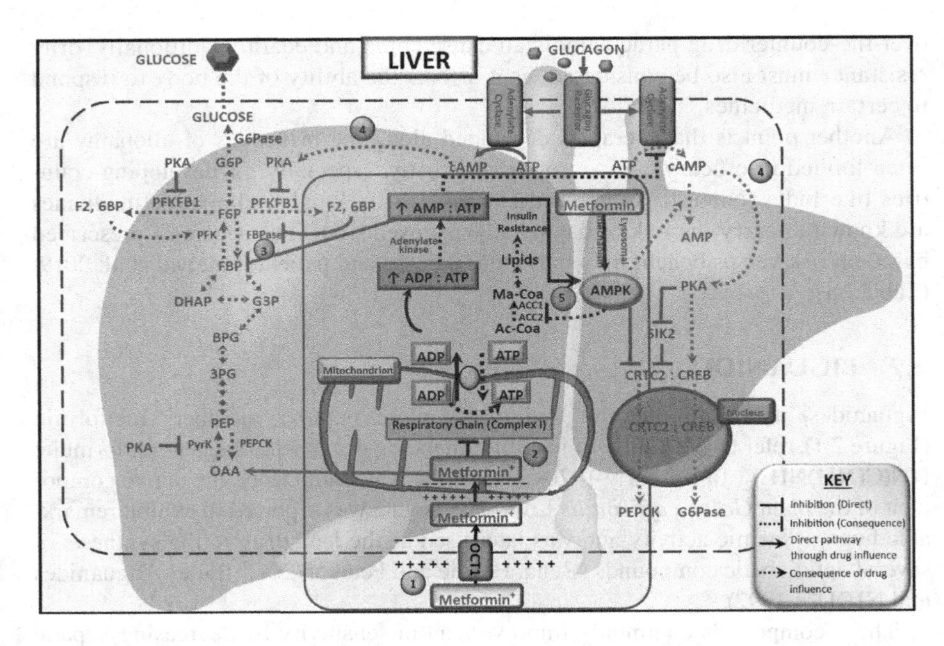

FIGURE 2.2 Schematic diagram of the primary sites of action of biguanides in the liver and the consequence of those actions. (The greyed areas of the diagram represent halted processes due to precursor inhibition.)

ratio, is the restriction in production of the protein kinase-A (PKA) enzyme (Step 4). PKA is known to be linked to both hepatic glucose production and the transcription of the genes for phosphoenolpyruvate carboxykinase (PEPCK) (an enzyme in the lyase family that is used in the metabolic pathway of gluconeogenesis) and glucose 6-phosphatase (G6Pase) (an enzyme that hydrolyzes glucose 6-phosphate, resulting in the creation of a phosphate group and free glucose). Finally, the activated AMPK phosphorylates the isoforms of acetyl CoA carboxylase (ACC) (ACC1 and ACC2), thus inhibiting fat synthesis and promoting fat oxidation instead (Step 5) (Rena, Hardie and Pearson 2017).

The use of biguanides comes with associated risks, as it was reported that continued use of metformin was associated with gastrointestinal (GI) side effects. However, causation of the side effects remains inconclusive. Other common side effects that have been reported from metformin use include nausea, diarrhoea, and weight loss (Wang and Hoyte 2019).

Other examples of biguanides, such as phenformin (Figure 2.3a) and buformin (Figure 2.3b), were marketed in the United States in the 1950s, subsequently becoming available across Canada, the United Kingdom, and Australia. However, these were removed from the market in the 1970s because of the toxic events that they were associated with in the body (Bailey 1992, Wang and Hoyte 2019). Additionally, metformin toxicity has also been known to lead to hyperlactatemia and metabolic acidosis, though the level of incidence is very low due to its acceptable level of excretion (Wang and Hoyte 2019).

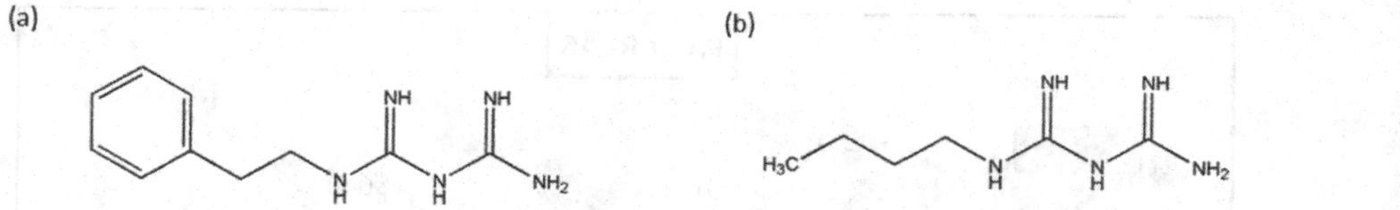

(a) (b)

FIGURE 2.3 Structures of examples of biguanides: (a) phenformin and (b) buformin.

In 2017, Abbas et al. formulated a new family of biguanide hydrochloride salts that lowered blood glucose (BG) levels in hyperglycaemic rats (Abbas et al. 2017, Basyouni et al. 2017). The antidiabetic activities of the novel salts were gauged against that of metformin and showed comparable and significant decreases in the elevated BG levels of the diabetic animals. As such, the results gave good justification for the compounds to enter clinical studies to assess their relative potentials as therapeutic agents in humans.

Finally, aside from their applications in DM therapy, in recent years scientists have started exploring alternative applications for the use of biguanides. Interestingly, great promise has been reported in their use as anticancer agents and this has encouraged further investigations in this area (Pollak 2013).

2.3 SULFONYLUREAS

Sulfonylureas are considered to be the "second-line treatment" after treatment failure with metformin (Douros, Dell'Aniello et al. 2018). Compounds belonging to this class of antidiabetic therapeutants possess an *S*-arylsulfonylurea structure, with a *p*-substituent on the phenyl ring (**R_1**) and various groups terminating the urea *N′* end group (**R_2**) (Figure 2.4).

The drugs function by stimulating the pancreas to release more insulin (Figure 2.5) (Douros, Dell'Aniello et al. 2018, Sola et al. 2015, Östenson et al. 1986), but this treatment is only effective if there is still residual pancreatic β-cell activity.

The intracellular mechanism of action in the pancreatic β-cells is initiated by the binding of the sulfonylureas to the K_{ATP} channels' SUR-1 receptors (Step 1). The binding causes the pancreatic ATP-sensitive potassium channels (KATP) to close, resulting in an elevated electrical potential and the subsequent depolarization of the β-cell's membrane (Step 2) (Douros, Dell'Aniello et al. 2018,

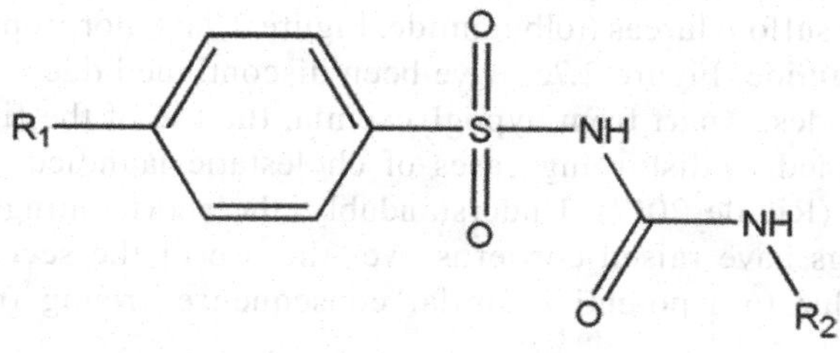

FIGURE 2.4 Structural formula of sulfonylureas.

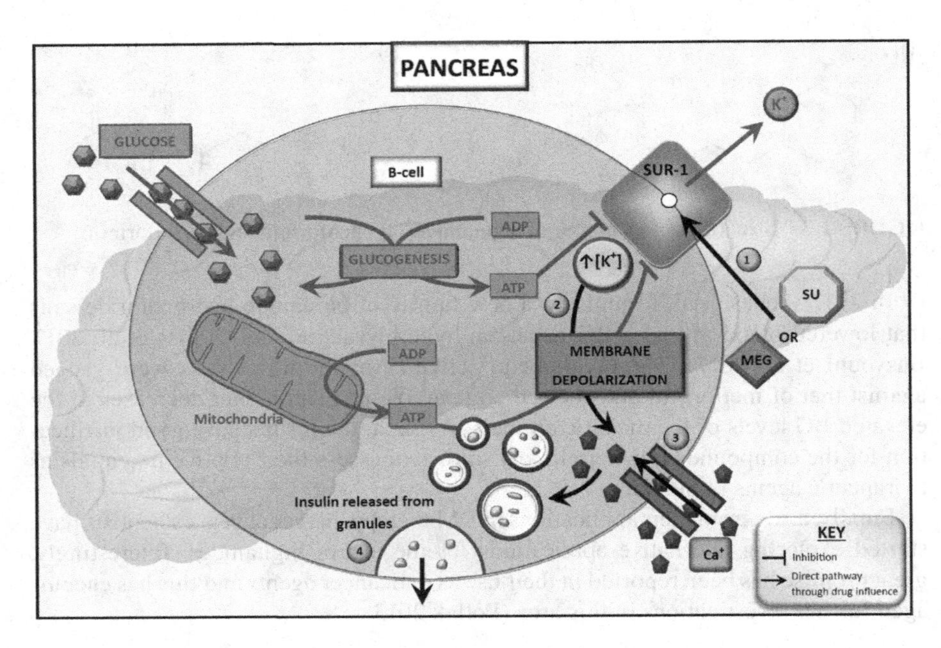

FIGURE 2.5 Schematic diagram of the effect of **sulfonylureas** and meglitinides in the pancreas.

Östenson et al. 1986). This depolarization leads to an influx of calcium ions (Ca^{2+}) into the cell (Step 3) which, in turn, triggers the insulin vesicles to release granulated insulin into the blood (Step 4). Apart from this, sulfonylureas have also been found to suppress glucagon secretion which, when in conjunction with the glucagon-like peptide-1 (GLP-1) (that is released from the small intestine), results in a decreased BG concentration (Östenson et al. 1986).

Sulfonylureas are classified into two "generations." First-generation drugs, which possess small, polar, water-soluble substitutions and their second-generation cousins, which have large, non-polar, and more lipid-soluble substitutions. The latter's pendant groups allow for the easier penetration of cell membranes, thus conferring a greater potency of insulin sensitivity and insulin secretion (R. K. Campbell 1998). Currently, the second-generation sulfonylureas (glyburide [glibenclamide: Figure 2.6a], glipizide [Figure 2.6b], glimepiride [Figure 2.6c], and gliclazide [Diamicron: Figure 2.6d]) continue to be used, while the use of first-generation sulfonylureas (tolbutamide: Figure 2.7a; chlorpropamide: Figure 2.7b; and tolazamide: Figure 2.7c) have been discontinued due to severe hypoglycaemic episodes. Apart from hypoglycaemia, the use of the first-generation drugs has also led to disturbing cases of cholestatic jaundice, hepatitis, and hepatic failure (Riddle 2017). Understandably, the shortcomings of the first-generation drugs have raised concerns over the use of the second-generation sulfonylureas due to a potential similar consequence arising from excessive dosage.

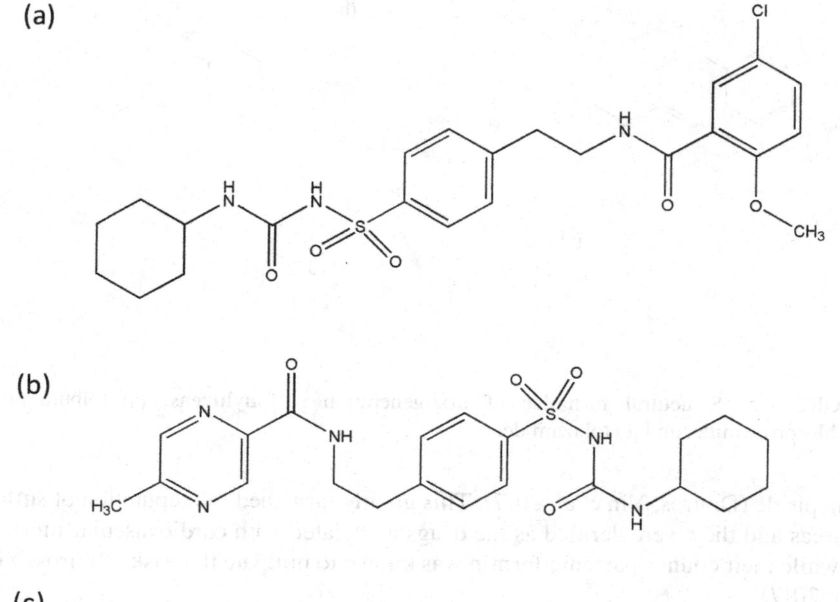

FIGURE 2.6 Structural formulae of second-generation **sulfonylureas**: (a) glibenclamide; (b) glipizide; (c) glimepiride, and (d) gliclazide.

The literature has shown that the use of tolbutamide has been associated with increased mortality in its patients (Sola et al. 2015, Riddle 2017). Douros et al. compiled several clinical and research studies concerning the association of sulfonylureas with the increased risk of cardiovascular events. They found that short-acting sulfonylureas – gliclazide, glipizide, and tolbutamide – could increase the risk of adverse cardiovascular events, compared to the long-acting sulfonylureas – glyburide and

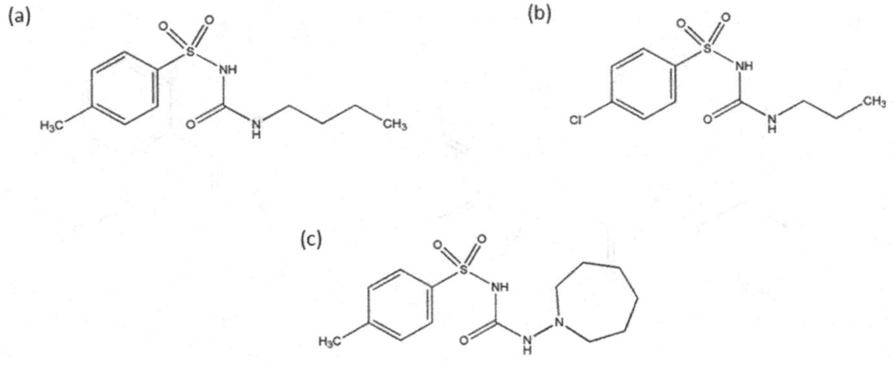

FIGURE 2.7 Structural formulae of first-generation **sulfonylureas**: (a) tolbutamide, (b) chlorpropamide, and (c) tolazamide.

glimepiride (Douros, Yin et al. 2017). This greatly tarnished the reputation of sulfonylureas and they were derided as the drugs associated with cardiovascular mortality, while their counterpart, metformin was known to mitigate that risk (Douros, Yin et al. 2017).

Despite the negativities, patients still continue to use this class of drug because of its relatively low cost, convenience in dosage to metformin, and reliability in action; with as little as one dose per day, sulfonylureas can attenuate spiked BG levels, with the only rare symptomatic side effects being hypoglycaemia and weight gain (Kalra, Madhu and Bajaj 2015).

2.4 MEGLITINIDES

Meglitinides are short-acting antidiabetic agents and are often called "non-sulfonylurea secretagogues" or "glinides." Meglitinides are the substrates of cytochrome P450 (CYP) enzymes and the organic anion transporting polypeptides 1B1 (OATP1B1 transporter) (Wu et al. 2017). Like sulfonylureas, they increase insulin secretion from the pancreas. However, their faster activity is compromised by their much shorter duration of action, which is mainly due to their relatively short half-lives. As illustrated in Figure 2.5, the mechanism of action of meglitinides is similar to that of sulfonylureas. Although seemingly identical, meglitinides differ from sulfonylureas in two ways: (a) the discussed signal is mediated through a different binding site on the SUR-1 receptor and (b) the differences in the drugs' physical and chemical characteristics.

In clinical practice, meglitinides possess a similar potency to metformin, and could serve as an alternative to metformin when the latter's side effects are intolerable (Wu et al. 2017, Diabetes Self-Management 2015). Like sulfonylureas, the most common side effects of meglitinides are hypoglycaemia and weight gain. However, the latter is still preferred due to its weaker binding to – and thus faster dissociation from – the receptor.

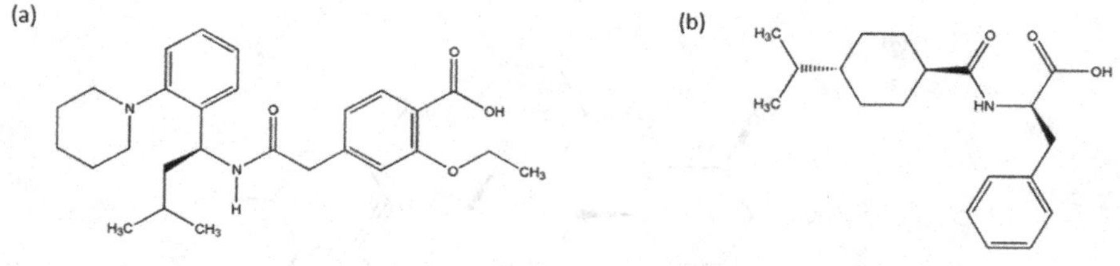

FIGURE 2.8 Structural formulae of the main brands of meglitinides: (a) repaglinide (Prandin) and (b) nateglinide (Starlix).

The main brands of this class of drugs are repaglinide (Figure 2.8a) and nateglinide (Figure 2.8b), both of which are metabolized in the liver, with <10% of repaglinide and most of nateglinide being excreted (Kawamori, Kaku et al. 2012).

Repaglinide was the first of the two meglitinides that were developed and used in adults with type-2 diabetes mellitus (T2DM) (Rosenstock et al. 2004). Like repaglinide, nateglinide also binds competitively to the sulfonylureas' receptor however, the pharmacodynamic properties of this molecule are unique in several aspects. Firstly, *in vitro* studies have indicated that nateglinide inhibits the potassium ATP channels faster than repaglinide and within a shorter duration of action (Guardado-Mendoza et al. 2013)., and the half-life of nateglinide on the SUR-1 receptor is approximately 5 seconds compared to that of repaglinide, which is approximately 3 minutes.

It was found that the duration of binding showed good correlation with relatively similar efficacy. In 2012, Kawamori et al. conducted a 16-week clinical trial aimed at investigating the relative efficacy and safety of repaglinide and nateglinide on 130 Japanese patients with T2DM and glycated haemoglobin (HbA1c). During the 16-week study period, the patients were treated with repaglinide (0.5 mg) and nateglinide (90 mg) three times a day together with diet and exercise. At the end of the study, the group reported that repaglinide monotherapy had given greater glycaemic improvement than nateglinide monotherapy, by more efficiently reducing both HbA1c and fasting plasma glucose (PG) values (Kawamori, Kaku et al. 2012).

In another study by Rosenstock et al., 150 T2DM patients were randomized to receive monotherapy of repaglinide (0.5 mg/meal, maximum dose 4 mg/meal) or nateglinide (60 mg/meal, maximum dose 120 mg/meal) for 16 weeks. Their report showed that both repaglinide and nateglinide had similar postprandial glycaemic effects on the patients, who were previously treated with diet and exercise. However, repaglinide monotherapy was found to be significantly more effective than nateglinide monotherapy in reducing HbA1c and fasting blood glucose (FBG) values, after 16 weeks of treatment (Rosenstock et al. 2004).

2.5 ALPHA-GLUCOSIDASE INHIBITORS

Alpha-glucosidase inhibitors (AGIs), such as acarbose (Glucobay) (Figure 2.9a), miglitol (Figure 2.9b), and voglibose (Figure 2.9c), form a new class of "suppressing-based" therapy that target the enzyme, α-glucosidase (AG). AG is an exo-type

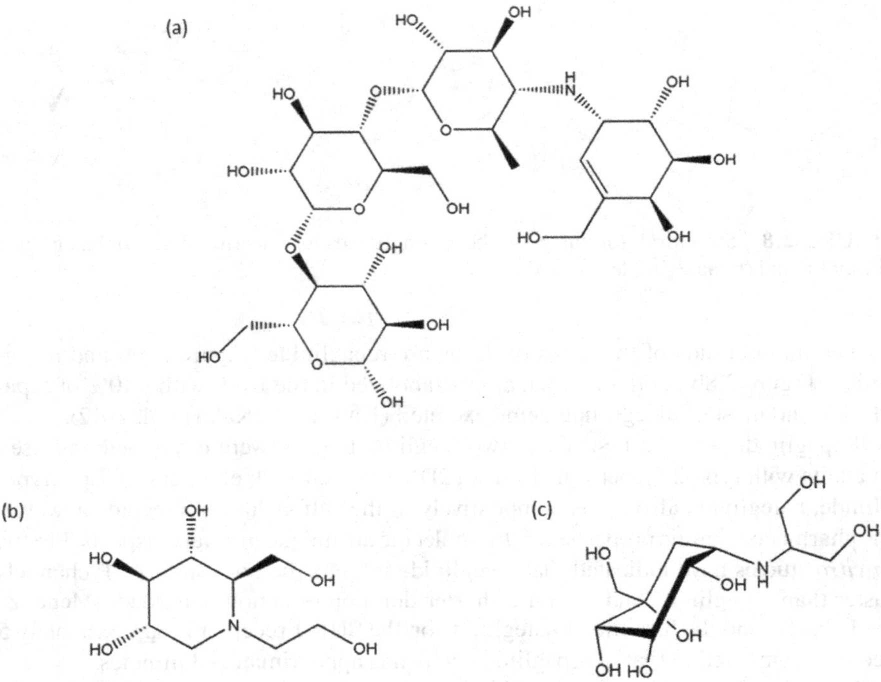

FIGURE 2.9 Structural formulae of α-glucosidase inhibitors: (a) acarbose, (b) miglitol, and (c) voglibose.

carbohydrase that is responsible for breaking down complex carbohydrates to glucose in the upper gastrointestinal tract (Kumar, Narwal et al. 2011, Lebovitz 1998). As such, the inhibition of the AG enzyme results in a decrease in complex carbohydrate catabolism, which as a consequence, retards the rise in postprandial BG levels (Lebovitz 1998).

As illustrated in Figure 2.10, by brushing the borders of the small intestine, the AG inhibitors competitively bind to the intestinal enzymes, thus hindering the formation of the enzyme–substrate complexes (Step 1). This leads to the slow absorption of carbohydrates (Step 2), which results in a greater amount of undigested and unabsorbed sugars (Step 3) (Dabhi, Bhatt and Shah 2013).

Acarbose is a pseudo-tetrasaccharide of microbial origin. Its chemical structure is analogous to that of oligosaccharides obtained from starch digestion and this morphological similarity allows for its competitive inhibitory action along the small intestine, up to the ileum (Rabasa-Lhoret and Chiasson 1998, Lin et al. 2003, Willms and Ruge 1999). Chiasson et al. reported on three *in vivo* studies that were conducted to compare the efficacy of acarbose treatments versus that of sulfonylureas. Using 255 diabetic patients, they found that the acarbose treatment resulted in an absolute reduction in HbA1c of 0.66 percentage points ± 0.28, compared to 0.88 points ± 0.28 for sulfonylurea treatment. This meant that acarbose was as effective as the sulfonylureas in the normalization of glucose levels (Leroux-Stewart, Rabasa-Lhoret and Chiasson 2015).

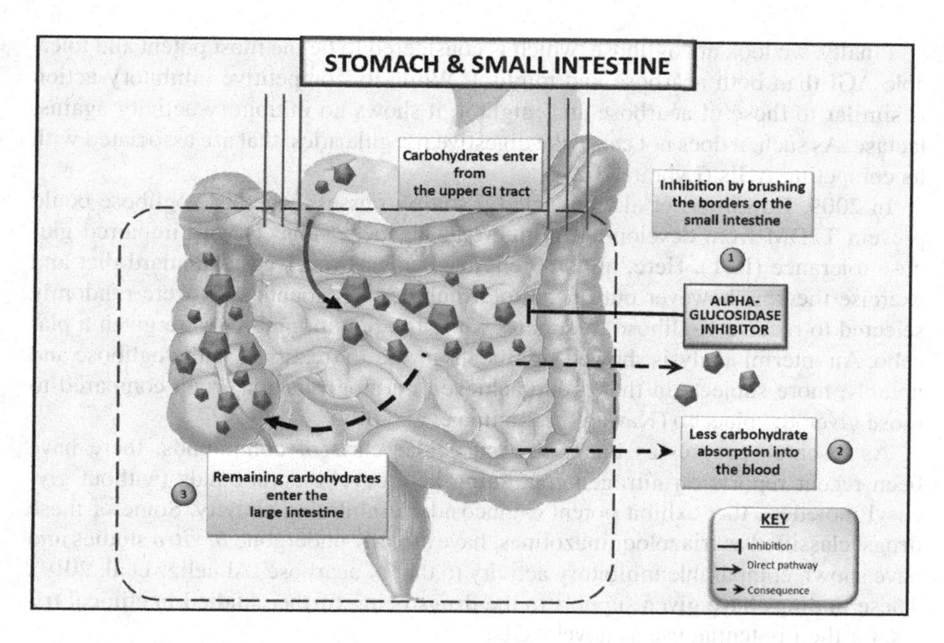

FIGURE 2.10 Schematic diagram of the effect of α-glucosidase inhibitors in the small intestine.

Miglitol is the first pseudo-monosaccharide that comes from a 1-deoxynojirimycin derivative (Kingma et al. 1992). It is excreted through the kidneys and unlike sulfonylureas, does not cause an increase in body weight and excessive hypoglycaemia, when administered as monotherapy (Scott and Spencer 2000). Furthermore, it is considered relatively safe, as it shows no significant effects on renal, cardiovascular, respiratory, or haematological parameters in long-term studies. Like acarbose, it improves glycaemic control, which is reflected in a reduced HbA1c level (Scott and Spencer 2000). However, unlike acarbose, miglitol is almost completely absorbed in the upper section of the small intestine.

Clinical trials conducted with the dosages of 50–100 mg of miglitol, three times a day in patients with T2DM have shown significant improvements in glycaemic control during a 6 to 12 month period (Scott and Spencer 2000, Campbell, Baker and Campbell 2000). Also, studies have shown that miglitol has a similar efficacy to that of acarbose (but at lower doses), and compared to sulfonylureas, miglitol provides a similar reduction in FBG and postprandial plasma glucose (PPG) levels (Kingma et al. 1992, Scott and Spencer 2000). As a common shortcoming however, both acarbose and miglitol have been reported to cause flatulence, abdominal pain, and diarrhoea.

In 2000, Scott and Spencer compared clinical trials composed of 1783 patients with T2DM, during a 6 to 12 month period. It was seen that some of the trials were insufficiently controlled by diet alone or diet plus sulfonylurea agents. However, one study using miglitol monotherapy of 50–100 mg, three times a day, resulted in a significant improvement in glycaemic control and a decrease in postprandial serum insulin levels (Kingma et al. 1992).

Finally, we look at voglibose, which is considered to be the most potent and tolerable AGI than both acarbose and miglitol. While its competitive inhibitory action is similar to those of acarbose and miglitol, it shows no inhibitory activity against lactase. As such, it does not cause the digestive irregularities, that are associated with its competing AGIs (Lybrate 2021).

In 2009, Kawamori et al. conducted a study to assess whether voglibose could prevent T2DM from developing in high-risk Japanese subjects with impaired glucose tolerance (IGT). Here, all 1780 eligible subjects received a standard diet and exercise therapy, however out of the total number of patients, 897 were randomly selected to receive voglibose treatment, while the remaining 883 were given a placebo. An interim analysis showed significant favour to treatment with voglibose and notably, more subjects in that group achieved normoglycaemia when compared to those given the placebo (Kawamori, Tajima et al. 2009).

As research progresses under this particular class of compounds, there have been recent reports on nitrogen-containing heterocyclic compounds (without glycosyl moieties), that exhibit potent α-glucosidase inhibiting activity. Some of these drugs, classified as triazoloquinazolines, have already undergone *in vitro* studies and have shown comparable inhibitory activity to that of acarbose (Abuelizz et al. 2019). These findings have given support to the drugs being further studied in clinical trials, for their potential use as novel AGIs.

2.6 DPP-4 INHIBITORS

Dipeptidyl-peptidase IV (DPP-4) inhibitors, such as sitagliptin (Januvia: Figure 2.11a) and vildagliptin (Galvus: Figure 2.11b), are a new class of incretin-based therapies. Also known as gliptins, they work by blocking the activity of DPP-4 enzymes, which are known to break down the insulin-secreting and glucagon-inhibiting incretins: glucagon-like peptide-1 (GLP-1) and glucose-dependent insulinotropic peptide (GIP) (Meier 2012).

Just like AG enzymes, DPP-4 enzymes are active in the small intestine. As illustrated in Figure 2.12, the DPP-4 inhibitor competitively binds to the DPP-4 enzymes (Step 1) and suppresses the enzyme's activity, thereby retarding the degradation rate of the incretins (Meier 2012). The incretins are therefore allowed to be released into the bloodstream (Step 2), where they act in the pancreas.

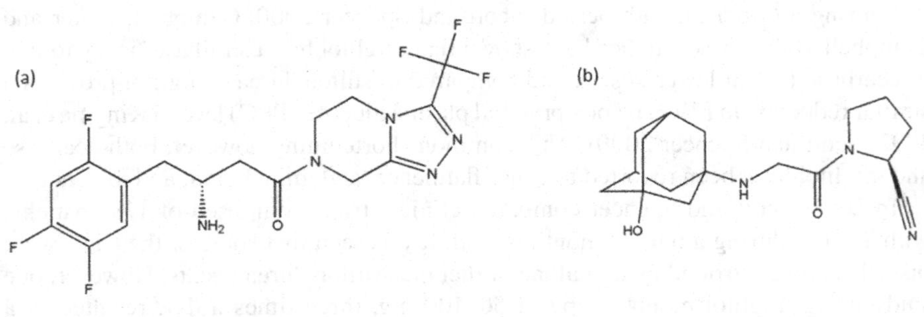

(a)

(b)

NH_2 O N N N N F F F F F F

N H O N HO O N

FIGURE 2.11 Structural formulae of dipeptidyl-peptidase IV (DPP-4) inhibitors: (a) sitagliptin and (b) vildagliptin.

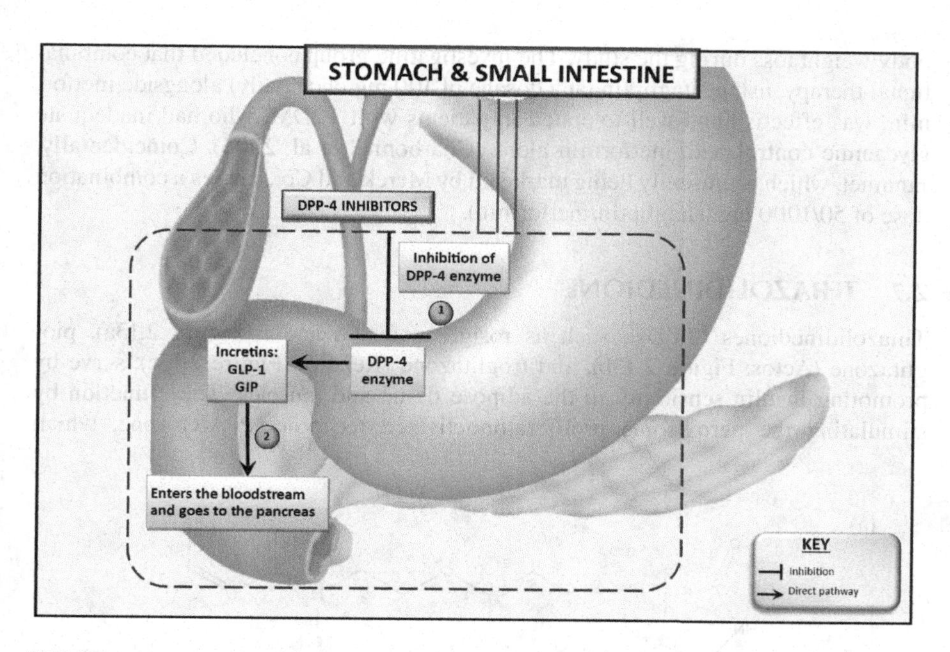

FIGURE 2.12 Schematic diagram of the effect of DPP-4 inhibitors in the small intestine.

Januvia was the first DPP-4 inhibitor approved by the US Food and Drug Administration (FDA) and marketed in the United States in 2006 for T2DM treatment. In 2010, patients who used Januvia for T2DM treatment filed a lawsuit against Merck & Co. with different law firms such as Morgan & Morgan, and The Drug Law Centre in the United States, because it was alleged that Januvia was linked to pancreatic cancer and pancreatitis. Close to 90 cases were issued against Merck & Co. On their website, Morgan & Morgan stated that Merck was facing several lawsuits, claiming that Januvia was a flawed and defective product, and was alleged to be associated with the risk of pancreatic cancer in its users. Further allegations stated that the company knew about this information beforehand and withheld the information from the public (Morgan & Morgan 2015). Despite this, the FDA chose to neither publish any warning of the associated pancreatic cancer risk of the drug, nor order a recall of Januvia from the market (Morgan & Morgan 2015, Drug Law Center 2017).

In a case study conducted by Charbonnel et al. in 2006, 701 patients with T2DM were screened and randomized to assess the efficacy of sitagliptin with metformin monotherapy. The patients, aged 19–78 years, had mild to moderate hyperglycaemia (mean A1C 8.0%) and were given doses of 1500 mg/day of metformin alongside a placebo or sitagliptin (100 mg once daily) in a 1 : 2 ratio for 24 weeks (Charbonnel et al. 2006, Gallwitz 2007). This combination resulted in sitagliptin treatment showing significant reductions compared with the placebo in A1C (–0.65%), FBG, and 2-hour post-meal glucose in some patients. A significantly greater proportion of the patients achieved an A1C of 7% with sitagliptin (47.0%) when compared to those receiving the placebo (18.3%), and did not experience any adverse effects from this treatment or increased risk of hypoglycaemia. However, some patients experienced body weight loss during the study. The investigating group concluded that combinational therapy, using sitagliptin (at a dosage of 100 mg once daily) alongside metformin, was effective and well-tolerated in patients with T2DM, who had inadequate glycaemic control with metformin alone (Charbonnel et al. 2006). Coincidentally, Janumet, which is presently being marketed by Merck and Co., carries a combination dose of 50/1000 mg (sitagliptin/metformin).

2.7 THIAZOLIDINEDIONE

Thiazolidinediones (TZDs), such as rosiglitazone (Avandia: Figure 2.13a), pioglitazone (Actos: Figure 2.13b), and troglitazone (Rezulin: Figure 2.13c), serve by promoting insulin sensitivity in the adipose tissue and muscles. They function by stimulating the peroxisome proliferator-activated receptor (PPAR) gene, which

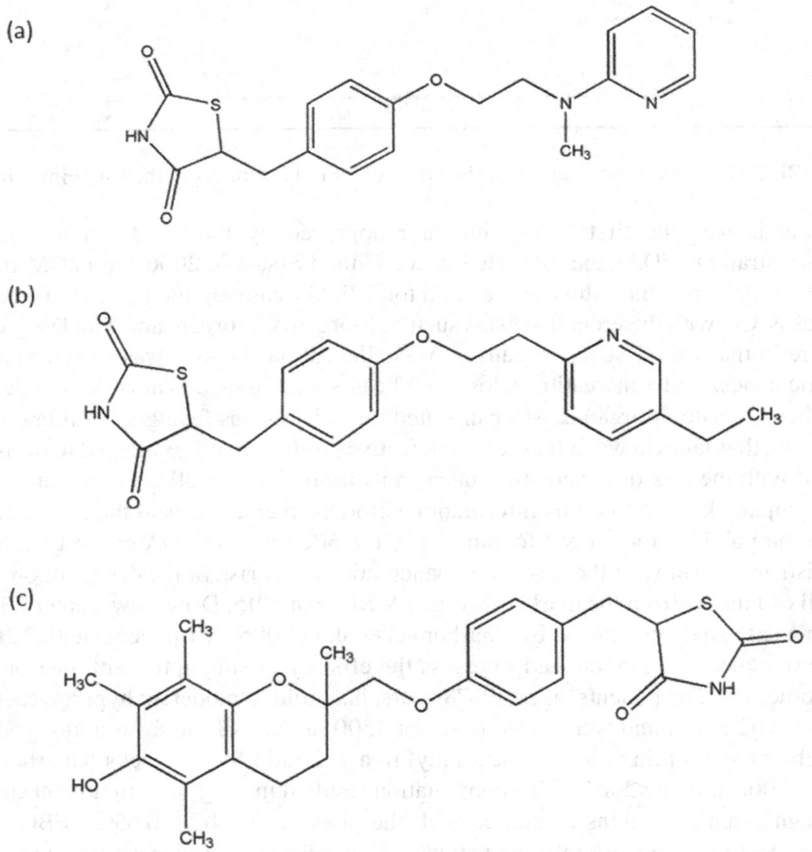

(a)

(b)

(c)

FIGURE 2.13 Structural formulae of thiazolidinediones: (a) rosiglitazone (Avandia), (b) pioglitazone (Actos), and (c) troglitazone (Rezulin).

influences insulin sensitization and enhances glucose metabolism. PPARs are ligand-activated transcription factors of the nuclear hormone receptor superfamily that is composed of the following three subtypes: PPARα, PPARγ, and PPARβ/δ (Tyagi et al. 2011, Wafer, Tandon and Minchin 2017, Kahn, Chen and Cohen 2000). As illustrated in Figure 2.14, TZDs enter the adipose tissue cell and activate the γ-gene (Step 1). Along with ATP (produced from glycolysis), these ligands together with the retinoid receptor X(RXR) form a nuclear receptor gene (Step 2), which undergoes transcription to mRNA (Step 3). Finally, the mRNA is used to signal the formation of vesicles for the purpose of insulin storage (Step 4) and transport across the tissue (Step 5) (Kahn, Chen and Cohen 2000).

A major side effect of thiazolidinediones, as exemplified by troglitazone (Rezulin), is liver toxicity. In 1997, troglitazone was removed from the UK market by its distributor Glaxo, because it was associated with an increased risk of liver failure (Graham et al. 2003). This later influenced the decision of the FDA to have the manufacturer remove the product from the US market for similar reasons. It was suggested that the mitochondrial dysfunction induced by troglitazone caused cytotoxicity in the liver, which led to liver failure and most times, death (Liao, Wang and Wong 2010).

In 2017, Wang et al. carried out a post-mortem comparison of several case studies investigating the efficacy and safety of thiazolidinediones in T2DM patients (Wang et al. 2017). While it was noted that the efficacy of the drugs was undisputed, the group underscored the negativities that could arise from their long-term use, citing that some trials reported hypoglycaemia and weight gain as adverse side effects.

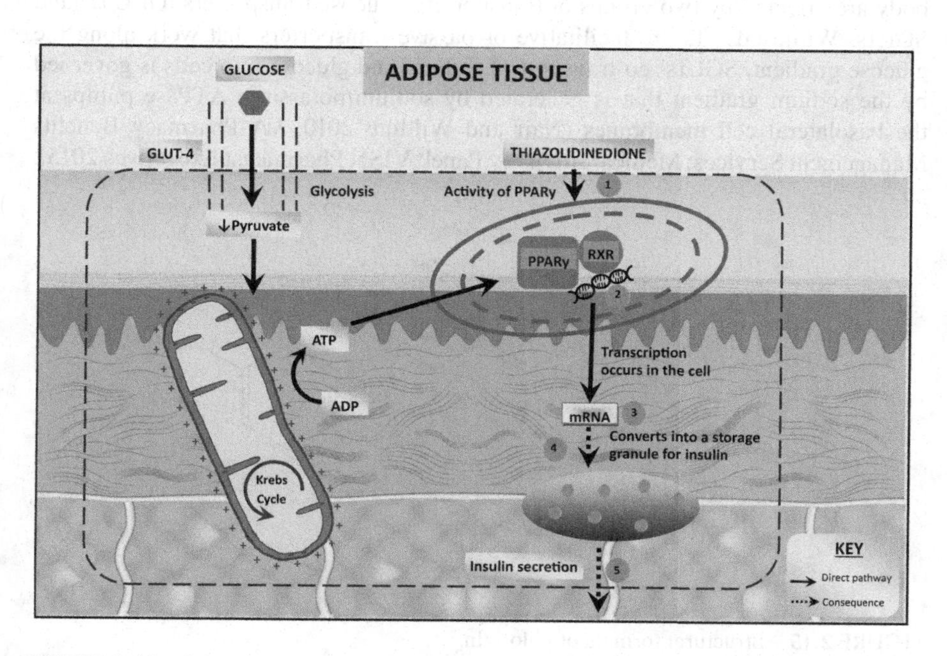

FIGURE 2.14 Schematic diagram of the effect of thiazolidinediones in adipose tissue.

2.8 SGLT-2 INHIBITORS

Recently, new treatments have emerged based on entirely new mechanisms. The story of the naturally extracted glucoside, phlorizin (Figure 2.15), proves to be an excellent case in point. In the 19th century, French chemists isolated the compound from the bark of apple trees and initially used it in the treatment of fever and infectious diseases (Nair and Wilding 2010). It was later found that its oral administration also led to the interference of carbohydrate absorption in the intestines, which resulted in glycosuria (Riddle and Cefalu 2018, Wikipedia 2021).

While further studies using the extract in diabetic animal models provided useful insights into the relationship between the reduction of hyperglycaemia and the restoration of insulin sensitivity, the drug was considered unsuitable for human use due to its undesirable side effects. However, its antidiabetic activity fuelled further investigations into its mechanism, which was traced to its interactions with sodium–glucose co-transporters (SGLTs) in the intestinal mucosa and the renal proximal tubule. This finding led to the development of newer drugs, which exhibited greater nuanced effects on these transporters (especially SGLT-2) within the kidney (Riddle and Cefalu 2018).

It is common knowledge to those in the field that the kidney plays a major role in glucose homeostasis through glomerular filtration and reabsorption of glucose (VA Pharmacy Benefits Management Services; Medical Advisory Panel; VISN Pharmacist Executives 2015). Glucose's absorption/reabsorption in the renal tubules, transport across the blood–brain barrier, and uptake and release by all cells in the body are effected by two groups of transporters: glucose transporters (GLUTs) and SGLTs. While GLUTs are facilitative or passive transporters that work along the glucose gradient, SGLTs' co-transport of sodium and glucose into cells is governed by the sodium gradient that is generated by sodium/potassium ATPase pumps at the basolateral cell membranes (Nair and Wilding 2010, VA Pharmacy Benefits Management Services; Medical Advisory Panel; VISN Pharmacist Executives 2015).

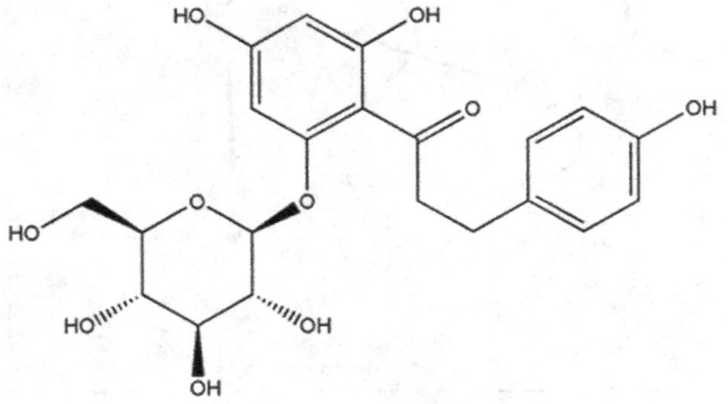

FIGURE 2.15 Structural formula of phlorizin.

It is known that renal tubular reabsorption undergoes adaptations in uncontrolled diabetes, particularly as it relates to the upregulation of renal GLUT vesicles. The increase in extracellular glucose concentration in diabetes lowers its outward-directed gradient from the tubular cells into the interstitium. As a natural response to maintain homeostasis, an upregulation of SGLT-2 is stimulated in an attempt to maintain renal tubular glucose reabsorption. This now manifests itself as a problem, as there is already a high BG level in the body. This situation (as illustrated in Figure 2.16) now presents itself as the ideal stage for the action of the SGLT2 inhibitors (SGLTi), where the drug's action results in the retardation of the rate of glucose reabsorption, consequently leading to a drop in the hyperglycaemic state (Nair and Wilding 2010).

Several specific and potent SGLTi drugs have undergone preclinical testing, most of which are glucosides that are structurally related to the lead drug: phlorizin. O-glucosides, such as sergliflozin etabonate (Figure 2.17a) and remogliflozin etabonate (Figure 2.17b), are strategically administered in the prodrug form of the biologically active sergliflozin A and remogliflozin to avoid degradation by β-glucosidase in the small intestine (Nair and Wilding 2010).

The clinical effects of the newer SGLTi drugs on T2DM include modest reductions in plasma glucose, A1C, body weight, and blood pressure, and are mediated in large part by glucosuria and sodium diuresis. These agents are taken once daily by mouth and do not require dose titration.

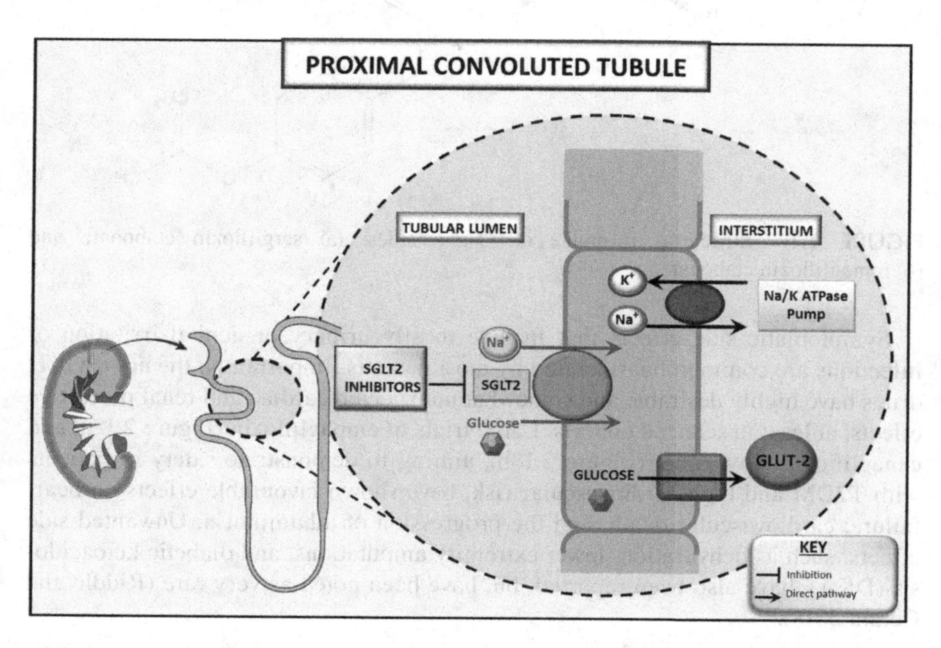

FIGURE 2.16 Mechanism of action for the inhibition of SGLT-2 in the S1 segment of the proximal convoluted tubule.

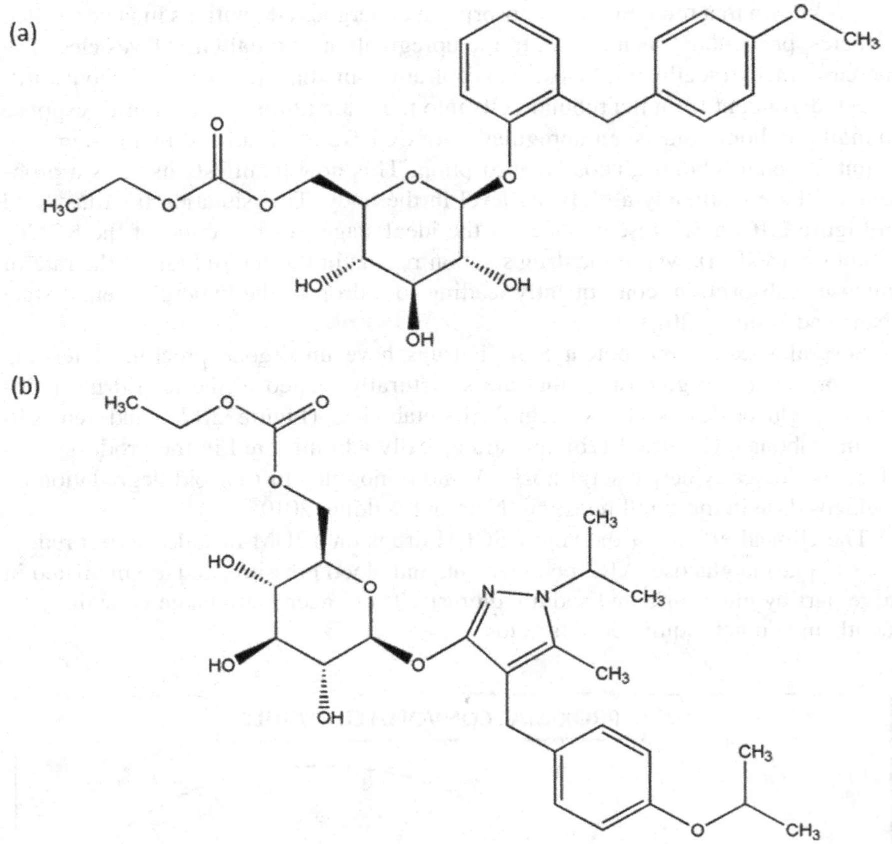

FIGURE 2.17 Structural formulae of O-glucosides: (a) sergliflozin etabonate and (b) remogliflozin etabonate.

Symptomatic side effects that include mostly urinary or genital irritation or infections are common but tolerated by most patients. Importantly, the new SGLTi drugs have highly desirable and somewhat unexpected cardiac and renal protective effects, at least in selected cohorts. Large trials of empagliflozin (Figure 2.18a) and canagliflozin (Invokana) (Figure 2.18b), aiming to demonstrate safety in patients with T2DM and high cardiovascular risk, have shown favourable effects on heart failure, cardiovascular death, and the progression of albuminuria. Unwanted side effects, such as dehydration, lower extremity amputations, and diabetic ketoacidosis (DKA), have also been reported, but have been noted as very rare (Riddle and Cefalu 2018).

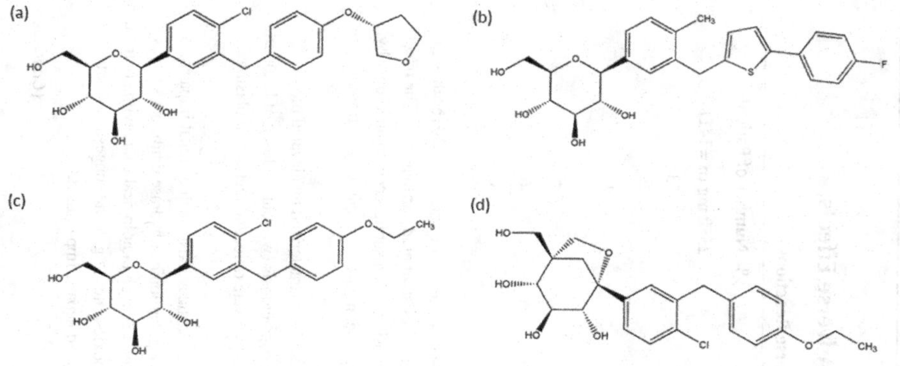

FIGURE 2.18 Structural formulae of SGLTi drugs: (a) empagliflozin, (b) canagliflozin, (c) dapagliflozin, and (d) ertugliflozin.

Although their underlying mechanisms are still not fully understood, the cardiovascular and renal benefits that have been achieved through the use of this class of drugs have warranted the great enthusiasm for its increased clinical use (Riddle and Cefalu 2018, Verma and McMurray 2018, Usman et al. 2018).

To date, SGLTi drugs present themselves as the new wonder drugs for T2DM therapy and the newer forms: empagliflozin, dapagliflozin (Figure 2.18c), and ertugliflozin (Figure 2.18d), which are being marketed under the brand names "Jardiance," "Farxiga," and "Steglatro," respectively, offer a greatly improved balance of benefits versus risks. Furthermore, ongoing studies aimed at assessing their effectiveness in T1DM therapy add more hope for the use of these drugs by a greater population of patients (Riddle and Cefalu 2018).

TABLE 2.1
Summary of Allopathic Preparations for T2DM Therapy, Mechanism of Actions, and Their Adverse Effects

Class; Marketed Name; Date of Entry	Mechanism of Action	Therapeutic Dosage (mg)	Adverse Reactions: Symptoms	Adverse Reactions: % Number of Patients
Biguanide:				**2550 mg (n = 141)**
Metformin (Glucophage) (Bristol-Myers 1995) (DrugLib .com 2009) *(First marketed in England in 1958; approved for use in Canada in 1972; and finally approved for use by the FDA in 1994)*	Decreases hepatic glucose production and intestinal absorption of glucose and increases peripheral glucose uptake and utilization	500 (twice daily) or 850 (once daily)	Diarrhoea	53
			Nausea/vomiting	26
			Flatulence	12
			Asthenia	9
			Indigestion	7
			Abdominal discomfort	6
			Headache	6
			Warning: Lactic Acidosis	
			Post-marketing cases of metformin-associated lactic acidosis have resulted in death, hypothermia, hypotension, and resistant brady arrhythmias. The onset of metformin-associated lactic acidosis is often subtle, accompanied only by non-specific symptoms such as malaise, myalgias, respiratory distress, somnolence, and abdominal pain.	
			Metformin-associated lactic acidosis was characterized by elevated blood lactate levels (>5 mmol/L), anion gap acidosis (without evidence of ketonuria or ketonemia), an increased lactate/pyruvate ratio, and metformin plasma levels generally >5 mcg/mL).	
			Risk factors for metformin-associated lactic acidosis include renal impairment, concomitant use of certain drugs (e.g. carbonic anhydrase inhibitors such as topiramate), age 65 years or greater, having a radiological study with contrast, surgery and other procedures, hypoxic states (e.g. acute congestive heart failure), excessive alcohol intake, and hepatic impairment.	

(Continued)

TABLE 2.1 (CONTINUED)
Summary of Allopathic Preparations for T2DM Therapy, Mechanism of Actions, and Their Adverse Effects

Class; Marketed Name; Date of Entry	Mechanism of Action	Therapeutic Dosage (mg)	Adverse Reactions	
			Symptoms	% Number of Patients
Biguanide:				
Phenformin (Information 2022) *(Entered the market in the 1970s, but was withdrawn in 1976 due to links to lactic acidosis)*	Decreases hepatic glucose production and intestinal absorption of glucose and increases peripheral glucose uptake and utilization	200	May cause metallic taste, nausea, anorexia, vomiting, diarrhoea, or cramps; weight loss sometimes occur.	
Sulfonylureas:				
Glyburide (Glibenclamide) (Diabeta, Bagomet Plus, Benimet, Glibomet, Gluconorm, Glucored, Glucovance, Metglib) (Sanofi-Aventis, DIABETA 2009) *(Entered the market in 1966, sold under different trade names)*	Stimulates the pancreas to produce more insulin	2.5–5	Gastrointestinal-related ailments	1.8
			Dermatological reactions	1.5
Sulfonylureas:				**(n = 702)**
Glipizide (Glucotrol) (Pfizer-Roerig 2008) *(Approved for use in 1971 and finally approved for use by the FDA in 1984)*	Stimulates the pancreas to produce more insulin	2.5–5	Combined observed ailments	11.8

(Continued)

TABLE 2.1 (CONTINUED)
Summary of Allopathic Preparations for T2DM Therapy, Mechanism of Actions, and Their Adverse Effects

Class; Marketed Name; Date of Entry	Mechanism of Action	Therapeutic Dosage (mg)	Adverse Reactions: Symptoms	Adverse Reactions: % Number of Patients	
Sulfonylureas:				(n = 746)	
Glimepiride (Amaryl) (Sanofi-Aventis, AMARYL 2013) *(Approved for use by the FDA in February 1999)*	Stimulates the pancreas to produce more insulin	1, 2, and 4	Dizziness	13	
			Asthenia	12	
			Headache	11	
			Nausea	8	
Sulfonylureas:				**30 mg (n=728)**	**80 mg (n=734)**
Gliclazide (Diamicron) (Servier 2018) *(Approved for medical use in 1972)*	Stimulates the pancreas to produce more insulin	30 and 80	Resistance mechanism disorder	8.8	6.4
			Respiratory disorder	21.4	20.9
			Musculo-skeletal disorders	15.2	14.6
			Secondary term	4.3	4.5
			Body as a whole	6	7.2
			Cardiovascular disease	6.5	7.3
			Urinary tract infection	2.6	3.0
			Gastrointestinal disorders	8.6	7.3
			Central peripheral nerve disorders	3.4	3.0
			Metabolic disorders	15.9	14.6
			Dermatological disorders	5.5	4.8
			Vision disorders	1.0	0.8
			Psychiatric disorders	3.0	3.2

(Continued)

TABLE 2.1 (CONTINUED)
Summary of Allopathic Preparations for T2DM Therapy, Mechanism of Actions, and Their Adverse Effects

Class; Marketed Name; Date of Entry	Mechanism of Action	Therapeutic Dosage (mg)	Adverse Reactions	
			Symptoms	% Number of Patients
Sulfonylureas:				**(n=352)**
Tolbutamide (Orinase) (L. Anderson et al. 2021) *(Entered the market in 1956, but was discontinued in the 1970s due to links to cardiovascular disease)*	Stimulates the pancreas to produce more insulin	500	Hypoglycaemia	–
			Gastrointestinal disorders	1–10
			Dermatological reactions	–
			Haematological reactions	0.1
			Metabolic reactions	–
			Endocrinological reactions	–
			Hypersensitivity	0.1
Sulfonylureas:				**(n=823)**
Tolazamide (Tolinase) (L. A. Anderson et al. 2022) *(Approved for use by the FDA in December 1987)*	Stimulates the pancreas to produce more insulin		Metabolic reactions	<0.01
			Gastrointestinal disorders	1–10
			Hypersensitivity	0.4
			Nerve-related reactions	0.1–1
			Hepatic reactions	0.1
			Haematological reactions	0.1

(Continued)

TABLE 2.1 (CONTINUED)
Summary of Allopathic Preparations for T2DM Therapy, Mechanism of Actions, and Their Adverse Effects

Class; Marketed Name; Date of Entry	Mechanism of Action	Therapeutic Dosage (mg)	Adverse Reactions	
			Symptoms	% Number of Patients
Meglitinide:				**(n=352)**
Repaglinide (Prandin) (Nordisk 2017) *(Approved for use by the FDA in December 1997)*	Stimulates the pancreas to produce more insulin	0.5, 1, and 2	Upper respiratory tract infection	16
			Headache	11
			Sinusitis	6
			Arthralgia	6
			Nausea	5
			Diarrhoea	5
			Back pain	5
			Rhinitis	3
			Constipation	3
			Vomiting	3
			Paraesthesia	3
			Chest pain	3
			Bronchitis	2
			Dyspepsia	2
			Urinary tract infection	2
			Tooth disorder	2
			Allergy	2
				(n = 1228)
			Hypoglycaemia	31
			Serious cardiovascular (CV) events	4
			Cardiac ischemic events	2
			Deaths due to CV events	0.5

(Continued)

TABLE 2.1 (CONTINUED)

Summary of Allopathic Preparations for T2DM Therapy, Mechanism of Actions, and Their Adverse Effects

Class; Marketed Name; Date of Entry	Mechanism of Action	Therapeutic Dosage (mg)	Adverse Reactions	
			Symptoms	% Number of Patients
Meglitinide:				**(n=352)**
Nateglinide (Starlix) (Cooperation 2017) *(Approved for use by the FDA in December 2000)*	Stimulates the pancreas to produce more insulin	60 and 120	Upper respiratory tract infection	16
			Headache	11
			Sinusitis	6
			Arthralgia	6
			Nausea	5
			Diarrhoea	5
			Back pain	5
			Rhinitis	3
			Constipation	3
			Vomiting	3
			Paraesthesia	3
			Chest pain	3
			Bronchitis	2
			Dyspepsia	2
			Urinary tract infection	2
			Tooth disorder	2
			Allergy	2
				(n = 1228)
			Hypoglycaemia	31
			Serious cardiovascular (CV) events	4
			Cardiac ischemic events	2
			Deaths due to CV events	0.5

(Continued)

TABLE 2.1 (CONTINUED)
Summary of Allopathic Preparations for T2DM Therapy, Mechanism of Actions, and Their Adverse Effects

Class; Marketed Name; Date of Entry	Mechanism of Action	Therapeutic Dosage (mg)	Adverse Reactions	
			Symptoms	% Number of Patients
Meglitinide:				**(n = 1441)**
Nateglinide (Starlix) (Cooperation 2017) *(Approved for use by the FDA in December 2000)*	Stimulates the pancreas to produce more insulin	60 and 120	Upper respiratory infection	10.5
			Back pain	4.0
			Flu symptoms	3.6
			Dizziness	3.6
			Arthropathy	3.3
			Diarrhoea	3.2
			Accidental trauma	2.9
			Bronchitis	2.7
			Coughing	2.4
			Non-severe hypoglycaemia	2.4
Alpha-glucosidase inhibitor:				**(n = 1255)**
Acarbose (Precose) (B. H. Pharmaceuticals 2011) *(Approved for use by the FDA in August 1999)*	Reduces the absorption of ingested carbohydrates	25, 50, and 100	Abdominal pain, diarrhoea, and flatulence	19, 31, and 74, respectively (at dosages of 50–300 mg)
Alpha-glucosidase inhibitor:				**(n = 962)**
Miglitol (Glyset) (B. H. Pharmaceuticals 2012) *(Approved for use by the FDA in August 1999)*	Reduces the absorption of ingested carbohydrates	25, 50, and 100	Abdominal pain, diarrhoea, and flatulence	11.7, 28.7, and 41.5, respectively (at dosages of 25 and 10 mg; three times daily)
			Dermatological reactions	4.3

(Continued)

TABLE 2.1 (CONTINUED)
Summary of Allopathic Preparations for T2DM Therapy, Mechanism of Actions, and Their Adverse Effects

Class; Marketed Name; Date of Entry	Mechanism of Action	Therapeutic Dosage (mg)	Adverse Reactions	
			Symptoms	% Number of Patients
Alpha-glucosidase inhibitor:				
Voglibose (Volicose) (Biocon 2022) *(Unconfirmed date of approval and entry into the market)*	Reduces the absorption of ingested carbohydrates	0.2 and 0.3	Gastrointestinal adverse effects such as diarrhoea, loose stools, abdominal pain, constipation, anorexia, nausea, vomiting, or heartburn may occur with the use of voglibose. Also abdominal distention, increased flatus, and intestinal obstruction–like symptoms due to an increase in intestinal gas.	
DPP-4 inhibitor:				**(n=443) @ 100 mg**
Sitagliptin (Januvia) (Incorporated 2021) *(Approved for use by the FDA in October 2006)*	Intensifies the effect of intestinal hormones (incretines) involved in the control of blood sugar	25, 50, and 100	Nasopharyngitis	5.2
				(n=179) @ 100 mg
			Upper respiratory infection	5.1
			Headache	2.8
DPP-4 inhibitor:				**(n=1855)**
Vildagliptin (Galvus) (Panacea 2019) *(Awaiting FDA approval)*	Intensifies the effect of intestinal hormones (incretines) involved in the control of blood sugar	50	Infections and infestations	Very rare (upper respiratory tract infection and nasopharyngitis)
			Metabolism and nutrition disorders	Uncommon (hypoglycaemia)
			Nervous system disorders	Common (dizziness) Uncommon (headache)
			Vascular disorders	Uncommon (peripheral oedema)
			Gastrointestinal disorders	Uncommon (constipation)
			Musculoskeletal and connective tissue disorders	

(Continued)

TABLE 2.1 (CONTINUED)
Summary of Allopathic Preparations for T2DM Therapy, Mechanism of Actions, and Their Adverse Effects

Class; Marketed Name; Date of Entry	Mechanism of Action	Therapeutic Dosage (mg)	Adverse Reactions	
			Symptoms	% Number of Patients
Thiazolidinedione:				**(n=2526)**
Rosiglitazone (Avandia) (GlaxoSmithKline 2007) *(Approved for use by the FDA in May 1999)*	Promotes insulin sensitivity in the adipose tissue and muscles	8	Upper respiratory tract infection	9.9
			Injury	7.6
			Headache	5.9
			Back pain	4.0
			Hyperglycaemia	3.9
			Fatigue	3.6
			Sinusitis	3.2
			Diarrhoea	2.3
			Hypoglycaemia	0.6
Thiazolidinedione:				**(n=2500)**
Pioglitazone (Actos) (Takeda Pharmaceuticals America 2011) *(Approved for use by the FDA in July 1999)*	Promotes insulin sensitivity in the adipose tissue and muscles	7.5, 15, 30, and 45	Upper respiratory tract infection	13.2
			Sinusitis	6.3
			Myalgia	5.4
			Tooth disorder	5.3
			Headache	9.1
			Pharyngitis	5.1
			DM aggravated	5.1

(Continued)

TABLE 2.1 (CONTINUED)
Summary of Allopathic Preparations for T2DM Therapy, Mechanism of Actions, and Their Adverse Effects

Class; Marketed Name; Date of Entry	Mechanism of Action	Therapeutic Dosage (mg)	Adverse Reactions	
			Symptoms	% Number of Patients (n=1450)
Thiazolidinedione:				
Troglitazone (Rezulin) (Limited 1999) *(Approved for use by the FDA in September 1999, but withdrawn from the market in 2000, due to its connection to five post-marketing cases of severe liver disease that resulted in death or liver transplantation)*	Promotes insulin sensitivity in the adipose tissue and muscles	200	Infection	18
			Pain	10
			Back pain	6
			Dizziness	6
			Asthenia	6
			Headache	11
			Nausea	6
			Urinary tract infection	5
			Peripheral edema	5
			Pharyngitis	5
			Rhinitis	5
			Backpain	6
			*During all clinical studies in North America, a total of 48 of 2510 (1.9%) Rezulin-treated patients and 3 of 475 (0.6%) placebo-treated patients had alanine transaminase (ALT) levels greater than three times the upper limit of normal (ULN). Nineteen patients (0.8%) had ALT over 8 × ULN, five patients (0.2%) had ALT values over 30 × ULN. Twenty of the Rezulin-treated and one of the placebo-treated patients were withdrawn from treatment. Two of the 20 Rezulin-treated patients developed reversible jaundice; one of these patients had a liver biopsy which was consistent with an idiosyncratic drug reaction. An additional Rezulin-treated patient had a liver biopsy which was also consistent with an idiosyncratic drug reaction.	

(Continued)

TABLE 2.1 (CONTINUED)
Summary of Allopathic Preparations for T2DM Therapy, Mechanism of Actions, and Their Adverse Effects

Class; Marketed Name; Date of Entry	Mechanism of Action	Therapeutic Dosage (mg)	Adverse Reactions: Symptoms	% Number of Patients: 100 mg (n = 833)	% Number of Patients: 300 mg (n = 834)
Sodium glucose co-transporter-2 (SGLT-2) inhibitor:					
Canagliflozin (Invokana) (Janssen Pharmaceuticals 2018) *(Approved for use by the FDA in September 2019)*	Helps the elimination of glucose via urine	100	Urinary tract infections*	5.1	4.6
			Increased urination**	2.9	3.8
			Thirst	2.8	2.4
			Constipation	1.8	2.4
			Nausea	2.1	2.3
				(n = 425)	**(n = 430)**
			Female genital mycotic infections	10.6	11.6
			Vulvovaginal pruritis	1.6	3.2
				(n = 408)	**(n = 404)**
			Male genital mycotic infections***	4.2	3.8

* Female genital mycotic infections include the following adverse reactions: Vulvovaginal candidiasis, vulvovaginal mycotic infection, vulvovaginitis, vaginal infection, vulvitis, and genital infection fungal.

**Increased urination includes the following adverse reactions: polyuria, pollakiuria, urine output increased, micturition urgency, and nocturia.

*** Male genital mycotic infections include the following adverse reactions: balanitis or balanoposthitis, balanitis candida, and genital fungal infections.

(Continued)

TABLE 2.1 (CONTINUED)
Summary of Allopathic Preparations for T2DM Therapy, Mechanism of Actions, and Their Adverse Effects

Class; Marketed Name; Date of Entry	Mechanism of Action	Therapeutic Dosage (mg)	Adverse Reactions		
			Symptoms	% Number of Patients	
				5 mg (n = 1145)	10 mg (n = 1193)
Sodium glucose co-transporter-2 (SGLT-2) inhibitor:					
Dapagliflozin (Farxiga) (A. Pharmaceuticals 2020) *(Approved for use by the FDA in January 2014)*	Helps the elimination of glucose via urine	10	Female genital mycotic infections	8.4	6.9
			Nasopharyngitis	6.6	6.3
			Urinary tract infections	5.7	4.3
			Back pain	3.4	4.2
			Increased urination	2.9	3.8
			Male genital mycotic infections	2.8	2.7
			Nausea	2.8	2.5
			Influenza	2.7	2.3
			Dyslipidaemia	2.1	2.5
			Constipation	2.2	1.9
			Discomfort with urination	1.6	2.1
			Pain in extremity	2.0	1.7

(Continued)

TABLE 2.1 (CONTINUED)
Summary of Allopathic Preparations for T2DM Therapy, Mechanism of Actions, and Their Adverse Effects

Class; Marketed Name; Date of Entry	Mechanism of Action	Therapeutic Dosage (mg)	Adverse Reactions		
			Symptoms	% Number of Patients	
Sodium glucose co-transporter-2 (SGLT-2) inhibitor:				**10 mg (n = 1145)**	**25 mg (n = 1193)**
Empagliflozin (Jardiance) (l. B. Pharmaceuticals 2022) *(Approved for use by the FDA in August 2014)*	Helps the elimination of glucose via urine	10 and 25	Urinary tract infections	9.3	7.6
			Female genital mycotic infections	5.4	6.4
			Upper respiratory tract infections	3.1	4.0
			Increased urination	3.4	3.2
			Dyslipidaemia	3.9	2.9
			Arthralgia	2.4	2.3
			Male genital mycotic infections	3.1	1.6
			Nausea	2.3	1.1
Sodium glucose co-transporter-2 (SGLT-2) inhibitor:				**5 mg (n = 519)**	**15 mg (n = 510)**
Ertugliflozin (Steglatro) (Inc. 2021) *(Approved for use by the FDA in December 2017)*	Helps the elimination of glucose via urine	5 and 15	Female genital mycotic infections	9.1	12.2
			Male genital mycotic infections	3.7	4.2
			Urinary tract infections	4.0	4.1
			Headache	3.5	2.9
			Increased urination	2.8	2.4
			Vaginal puritis	2.7	2.4
			Nasopharyngitis	2.5	2.0
			Back pain	1.7	2.5
			Weight decrease	1.2	2.4
			Thirst	2.7	1.4

3 Ayurvedic Medicines (*Ethnopharmacological Treatments*)

They sent Me to tell You that We are Truly Sorry...
You were Right and We were Wrong...
They Really do Work!

Modern science has tested the many claims of ethnomedicine and the majority of studies have shown them to be true. However, the possibility of drug–drug interactions or the occurrence of cytotoxic events due to wrongful dosages remain a concern. Thankfully, progressive research continues to churn out the solutions to these apprehensions.

3.1 INTRODUCTION

3.1.1 Overview and Significance

Apart from the commonly recognized allopathic drugs that are used in the treatment of diseases, a greatly regarded source of medicine throughout human history is natural herbs, and their worldwide popularity in present medicinal therapies indicates that they have grown in acceptance in modern medicine. Correspondingly, the use of herbal remedies is gaining more significance owing to the drawbacks and limitations noted with many allopathic drugs (Table 3.1).

DOI: 10.1201/9781003322429-3

Esteemed researchers in the field of ethnomedicine such as Associate Professor, Dr Amala Soumyanath (Wright et al. 2022), as well as corporations such as Shaman Pharmaceuticals Inc. (Reed and Scribner 1999, Oubré et al. 1997) have added to creating a stable foundation for further Ayurvedic medicines to be investigated and accepted.

3.1.2 History

Traditional medicine is a bank of knowledge, skills, and practices that, whether justifiable or not, is used in the prevention, diagnosis, and treatment of both physical and mental illnesses. When referring to traditional medicine outside its regular culture it is often referred to as "complementary" or "alternative" medicine. The three major traditional medicines that are globally recognized are Ayurveda or Traditional Indian Medicine (TIM) in India, Traditional Chinese Medicine (TCM) in China, and Traditional Arabic and Islamic Medicine (TAIM) in the Middle East.

India and China, whose traditional therapies are highly complementary, have historical origins of mutual learning and development, from medical theory to drugs used. The exchange of traditional medicines between China and India began in the Qin and Han dynasties (221 BC–220 AD), prospered in the Tang dynasty (618–907 AD), and declined after the Song dynasty (960–1279 AD) (Wu, Chen and Wang 2021).

TCM intellectuals are yet to gain a thorough understanding of the origin and historical development of TIM, but it is this understanding that will allow the joint promotion of traditional medicines from both countries.

3.1.2.1 Traditional Indian Medicine

In India, it is common medicinal practice for practitioners to formulate and dispense their own plant-based concoctions, using medicinal plants that have been traditionally used for over 1000 years. Prior to 1970, it was estimated that approximately 70% of the Indian population relied on Ayurvedic medicine since the national healthcare structure at the time could only provide for about 30% of the 1.38 billion population (Shi, Zhang and Li 2021).

During the Vedic period (10th century BC) in ancient India, one of the oldest medical systems (the Ayurvedic system) was developed into a comprehensive medical structure, which is now recognized as the AYUSH system. AYUSH has become a popular acronym that is used to describe TIM; it stands for Ayurveda, yoga and naturopathy, Unani, siddha, Sowa Rigpa, and homoeopathy (Wu, Chen and Wang 2021).

India attaches great importance to the management of, education in, and industry of traditional medicine, and has made various efforts to protect intellectual property (IP) rights. As such, it established the Ministry of AYUSH of the Government of India, which is responsible for traditional medicine in the country. Furthermore, a "Traditional Knowledge Digital Library" and special investment foundation were formed to archive traditional medical knowledge and to promote the industrialization of TIM. Regarding the IP rights, the National Biodiversity Authority must grant consent for research based on Indian biological resources or related traditional knowledge (Wu, Chen and Wang 2021). Additionally, the Indian Pharmacopoeia Committee and the Indian Laboratory Committee execute the standardization and modernization of Indian medicine, and specifically organize the preparation and printing of publications such as the Traditional Indian Pharmacopoeia (Shi, Zhang and Li 2021).

In 2002, the state introduced policies on the Indian medical system, with the main objectives to (1) use Ayurveda to promote good health and to promote healthcare for the local people (mainly those who cannot afford or do not have access to modern medical facilities) by means of prevention, promotion, mitigation, and treatment; (2) provide affordable, safe, and effective Ayurvedic services and medicines; (3) ensure that the supply, and authentic products of active pharmaceutical ingredients (APIs) meet the pharmacopoeia standard requirements, to help improve the quality of drugs for domestic use and/or export; (4) integrate Ayurveda into the medical service system and national planning, and ensure the maximum possible use of the huge infrastructure of hospitals, pharmacies, and doctors; and (5) provide full opportunities for the expansion and development of Indian pharmaceutical systems (ISM), utilize their potential, and revive their glory (Shi, Zhang and Li 2021).

3.1.2.2 Traditional Chinese Medicine

Traditional Chinese medicine is a broad range of medicine practices that share common concepts, and an integrated theory system. It has progressed through over 3000 years of clinical and pharmacological trials, as one of the longest medical practices in China and the Asia-Pacific areas. In China, TCM is considered equal to that of Western medicine, and there has been an integration between these two in the Chinese healthcare systems. However, TCM still lacks international recognition because of the scarcity of systemic research and evidenced investigation.

TCM has strong associations with yin-yang theory; five elements theory; the concept of qi; internal organ systems; and other vital systems such as blood, essence, and fluids. TCM also includes ethnic medicine (e.g. Tibetan medicine) and religious medicine (e.g. Buddhist medicine). It further and more specifically includes various forms of acupuncture, dietary therapy, herbal medicine, moxibustion, and physical exercise, which collectively predate to the birth of Chinese civilization (Wu, Chen and Wang 2021, W. Wang 2016).

Similar to the Ministry of AYUSH of the Government of India, the State Administration of Traditional Chinese Medicine oversees traditional medicine in the country. By the end of 2018, there were 60,700 Chinese medicine, medical, and health institutions with over 1.2 million beds, 715,000 personnel, over 1 billion people diagnosed and treated, and an estimated 36 million discharged. The progress of TCM can further be seen by the number of TCM pharmaceutical enterprises, colleges and universities of TCM, and non-Chinese medicine colleges and universities offering TCM, with a high student intake (Wu, Chen and Wang 2021).

In 2016, it was reported that the market scale of the TCM health industry totalled 1.75 trillion yuan; in 2018, the import and export volume of TCM products was reported at over 36 billion yuan; and in 2017, the annual trade volume of the industry was almost 500 billion yuan. Furthermore, there are more than 300,000 TCM clinics worldwide, and approximately 4 billion people use Chinese herbal medicine products (Wu, Chen and Wang 2021).

3.1.2.3 Traditional Arabic and Islamic Medicine

Herbal medicine has also been found as part of modern healing in the Middle East. More specifically referred to as Traditional Arabic and Islamic Medicine, it has grown in acceptance, interest, and respect by many of those in the herbal and scientific communities. TAIM refers to healing practices, beliefs, and philosophy incorporating herbal medicines, spiritual therapies, dietary practices, mind–body practices, and manual techniques, which may be applied singularly or in combination to treat, diagnose, and prevent illnesses and/or maintain well-being (Al Rawi et al. 2017).

TAIM came into existence hundreds of years ago, and investigations into its associated herbs have been conducted in many Arab countries such as Syria, Morocco, Yemen, and Egypt. The most recent survey conducted to investigate the possible uses of Mediterranean plants recorded many plant species belonging to different families still in use; however, the number is actually less than the over 700 known species of plants that were previously recorded as medicinal plants because of several factors endangering plant diversity or even causing the eradication of these herbs (Al Rawi et al. 2017, Azaizeh et al. 2010).

TAIM is the first choice for many in dealing with ailments such as infertility, epilepsy, psychosomatic troubles, and depression, and has been found to be quite successful in treating both acute and chronic diseases. Typically, a decoction is prepared by boiling plant parts in hot water, infusing in water or oil, or inhaling essential oils. It is also taken as juice, syrup, roasted material, fresh salad or fruit, macerated plant parts, oil, milky sap, poultice, or paste. TAIM treatments have been tested in cooperation with physicians and are now routinely prescribed to patients in Europe and in Mediterranean countries (Azaizeh et al. 2010).

Similar to TIM and TCM, it is well documented that TAIM has significantly impacted the development of modern medicine in Europe and remains one of the closest forms of original European medicine. Also, studies indicate that the eastern region of the Mediterranean has been identified as having a rich inventory of natural medicinal herbs. In recent times, the Galilee Society Research and Development Center, in cooperation with different institutes, collected data, which indicated that 200–250 herbs are still used in human treatment, and are commercialized regionally and internationally (Azaizeh et al. 2010, Masic et al. 2017).

3.1.3 Downfalls

Although the advantages of ethnopharmacological and other natural treatments are significant, particularly from an economic perspective as well as their availability on a global scale, it is vital to also acknowledge the possible shortcomings of these forms of therapies.

Paudyal et al. recognized the importance of highlighting the possible adverse effects arising from the proclivities of Ayurvedic medicines, possibly due to adulteration or inherent constituents such as alkaloids. As such, proper advising by health professionals is crucial in minimizing harm (Paudyal et al. 2019). Additionally, it should be taken into consideration that there exists the common misconception that alternative treatments are harmless because they are natural and are sometimes promoted as completely safe; however, natural does not equate to harmless. Since allopathic medicines are known to have adverse effects and are prescribed based on a risk versus benefit basis, many persons assume that these remedies are devoid of adverse effects because similar information, regarding alternative treatments, is not as readily available to them.

3.2 *ABELMOSCHUS ESCULENTUS* LINN. *(AE)*

Common names: Bindi, Lady Fingers, Ochro, Okra, Okro (Figure 3.1)

A. esculentus is a perennial plant that is often cultivated as an annual in temperate climates. As a member of the Malvaceae family, it is related to species such as cotton, cocoa, and hibiscus. Its leaves are 10–20 cm long and broad, palmately lobed, with five to seven lobes. Its flowers are 4–8 cm in diameter, with five white to yellow petals, often with a red or purple spot at the base of each petal. Its fruit is a capsule up to 18 cm long (however other species are known to grow to over 30 cm), with a pentagonal cross section, containing numerous seeds (Wikipedia 2021).

It was initially proposed that the antidiabetic effect of AE be attributed to its high mucilage content, which was thought to delay the absorption of glucose into the bloodstream. However, this idea was debunked as it was later demonstrated that the

FIGURE 3.1 Illustration of *Abelmoschus esculentus*.

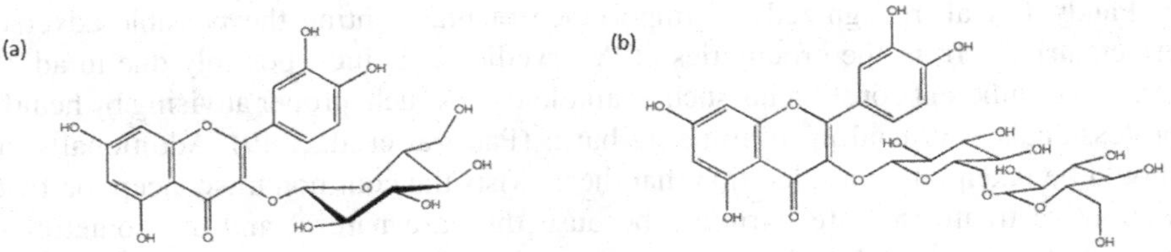

FIGURE 3.2 Structural formulae of isolated compounds from *Abelmoschus esculentus*: (a) isoquercetin and (b) quercetin-3-O-β- glucopyranosyl-(1→6)-glucoside.

mucilage yield, glucose entrapping capability, and α-glucosidase inhibitory potential of the aqueous fruit extract were relatively low. Interestingly, *in vivo* studies showed that it was the seeds of the fruit that actually conferred the plant's antidiabetic activity. A report suggested that the hypoglycaemic effect observed in rats, that were fed AE seed extract, was due to the inhibition of the intestinal α-glucosidase enzymes by two isolated flavonols: isoquercetin (Figure 3.2a) and quercetin-3-O-β-glucopyranosyl-(1→6)-glucoside (Figure 3.2b) (Thanakosai and Phuwapraisirisan 2013).

Another study by Sabitha and coworkers, investigated the antidiabetic and anti-hyperlipidemic potential of the peel (AEP) and seed (AES) powders in streptozotocin (STZ)-induced diabetic rats. Acute toxicity assays on both powders showed no signs of toxicity-related illnesses or death, at doses as high as 2000 mg/kg. From this, suitable therapeutic doses of AEP (100 mg/kg.bw) and AES (200 mg/kg.bw) were administered to the diabetic rats. The results showed a significant reduction in blood glucose (BG) levels and a simultaneous increase in the body weight of the treated subjects, relative to their untreated diabetic counterparts (Sabitha et al. 2011).

Alternatively, Herowati et al. used both aqueous and fractionated organic extracts of the AE fruit to assess its hypoglycaemic activity in insulin-resistant rats. It was reported that while the extract and its fractions showed antihyperglycaemic effects, they were still lower than that of the control standard, metformin. Additionally, the group observed and reported differences in the mechanisms of action due to the difference in the nature of the extracts. The presence of the water-favoured flavonols (isoquercetin and quercetin-3-O-β-glucopyranosyl-glucoside) acted as α-glucosidase inhibitors (AGI) and thus inhibited the absorption of maltase and sucrose in the intestine. However, the non-polar hexane fractions (containing non-polar compounds such as phytosterols and fatty acids) delayed the rate of gastric emptying, while the ethyl acetate fractions (containing alkaloids, terpenoids, and many flavonoids) positively influenced the translocation of the glucose transporter type-4 (GLUT-4) vesicles (Herowati et al. 2020).

In this next instance, Ramachandran et al. tested the antihyperglycaemic activity of the AE fruit extract in alloxan-induced diabetic rats. Here, 100 and 400 mg/kg .bw doses of extract were administered to both diabetic and non-diabetic subjects over a 15-day period. Subsequently, they observed a return to normoglycaemic levels for the treated diabetic subjects, suggesting that the constituent compounds were in some way either influencing an increase in glucose uptake or potentiating the secretion of insulin by the pancreas. The group further reported that continued treatment until Day 21 showed sustained control over the elevated BG levels of the diabetic rats, but not over the levels of the normal rats (Ramachandran et al. 2010).

In an interesting report by Khatun et al., aqueous AE fruit fractions were co-administered alongside metformin in an attempt to compare the effects of the absorption of orally administered glucose in Long–Evans rats. As expected, it was found that the extract significantly reduced the absorption of glucose in both 24-hour fasting rats and alloxan-induced diabetic rats. However, it was reported that the co-administration of the extract alongside metformin resulted in a drug–drug interaction that resulted in the near-complete loss of metformin's antihyperglycaemic activity. Little change was observed in the BG levels of the "double-treated" rats (33.5–32.2 mmol/L at 4 hours) in relation to the animals that received only the metformin therapy (a drop to 14.9 mmol/L within 4 hours) (Khatun et al. 2011).

Finally, Uraku et al. demonstrated the restorative potential of the AE fruit extract on several liver enzymatic activities. Using alloxan-induced diabetic rats, the group reported that elevated levels of alkaline phosphatase (ALP), aspartate aminotransferase (AST), and alanine aminotransferase (ALT) activities were attenuated in a dose-dependent manner, when the diabetic rats were treated with varying concentrations of the extract (Uraku et al. 2011).

3.3 *AGERATUM CONYZOIDES* LINN. (*AgC*)

Common names: Billygoat-weed, Tropical Whiteweed, Zeb-a-fam (Figure 3.3)

A. conyzoides is an annual herb that belongs to the Asteraceae family. It is mostly grown in Asia, temperate Brazil, and areas that are hot and humid. It can be identified by its shallow fibrous roots and colourful flowers, which can be either

FIGURE 3.3 Illustration of *Ageratum conyzoides.*

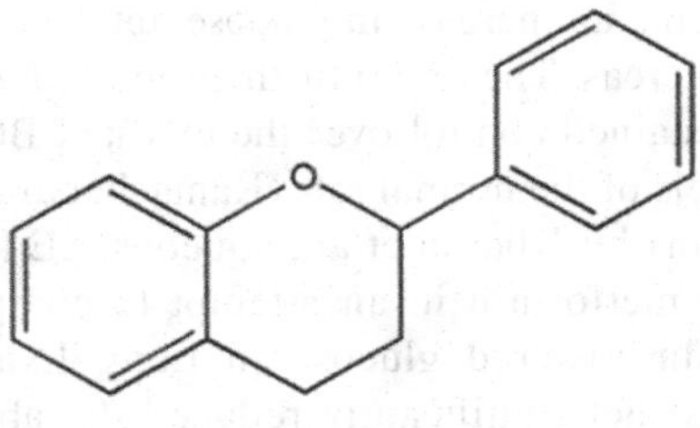

FIGURE 3.4 General structural formula of flavonoids (polyphenolic compounds).

purple, blue, or white (The Environmental Weeds of Australia n.d.). Its leaves, juices, and extracted essential oil (containing alkaloids, cardenolides, tannins, saponins, and flavonoids [Figure 3.4]) (Egunyomi, Gbadamosi and Animashahun 2011) are known to possess anti-inflammatory properties, as well as display antibacterial and antidiarrhoeal effects. However, studies have also shown it to exhibit antidiabetic activity.

Egunyomi et al. investigated the hypoglycaemic activity of an ethanolic extract of AgC shoots on alloxan-induced diabetic rats. The rats were divided into six groups: Group 1: negative control (healthy and untreated); Group 2: positive control (alloxan induced, untreated); Group 3: alloxan-induced diabetic rats (treated with 100 mg/kg.bw of the ethanol extract); Group 4: alloxan-induced diabetic rats (treated with 200 mg/kg.bw of the extract); Group 5: alloxan-induced diabetic rats (treated with 400 mg/kg.bw of the extract); and Group 6: normoglycaemic rats (healthy, but administered 400 mg/kg.bw of the extract). Appreciatively, it was reported that there was a significant decrease in fasting blood glucose (FBG) levels in Groups 3, 4, and 5 after the 2-week study period. In Group 3, the FBG levels decreased from 390.6 to 90.2 mg/dL; in Group 4, from 590.4 to 45.8 mg/dL; and in Group 5, from 466.2 to 42.4 mg/dL. Interestingly, in Group 6, the FBG levels only decreased in the last 2 days, from 55.0 to 49.3 mg/dL. Based on the *in vivo* experiment and phytochemical screenings results, the team was able to confirm the hypoglycaemic activity of AgC and thus its potential use as a phytomedicine for type-2 diabetes mellitus (T2DM) (Egunyomi, Gbadamosi and Animashahun 2011).

Nyunaï et al. also investigated the antihyperglycaemic properties of AgC in STZ-induced diabetic rats. Compared to Egunyomi et al.'s study, the present group used an aqueous extract from the leaves and administered three different doses, 100, 200, and 300 mg/kg.bw, for up to 8 hours. It was reported that both the 200 and 300 mg/kg.bw doses produced a significant reduction in the BG levels of the rats after 1.5 hours, up to the 8-hour stop time (Nyunaï, Njikam et al. 2009). Furthermore, the 300 mg/kg.bw dose of the extract produced the maximum reduction in the BG levels (36.85%), while 200 mg/kg.bw reduced the levels by 33% and 100 mg/kg.bw by 20.3% at the end of the 8-hour period. At the end, the group awarded greatest efficacy to the 200 mg/kg.bw dose, because of its consistency throughout the 8 hours.

In another report, Nyunaï and coworkers performed further studies, administering aqueous extract fractions: F1a, F1b, and F1c to STZ-induced diabetic rats at doses of 16 mg/kg.bw of F1a, 16 mg/kg.bw of F1b, and 88 mg/kg.bw of F1c. Although all the subfractions (F1a, F1b, and F1c) were found to be active against increases in BG, it was the F1c subfraction that had a significant effect. The group also indicated that the F1c fraction was more potent than the others because it showed similar activity compared to the control standard, glibenclamide (Nyunaï, Manguelle-Dicoum et al. 2010).

Finally, Agunbiade et al. investigated the hypoglycaemic effect of the aqueous leaf extract on alloxan-induced diabetic rats. Using a similar protocol to Nynai, they administered 500 mg/kg.bw of the AgC aqueous extract to the rats and compared it to the same dose of the commercial drug, glibenclamide. After treatment, the group reported a satisfactory reduction (by 39.1%) in the FBG levels of the rats; from a starting level of 309.6 mg/dL and then dropping to a level of 195.8 mg/dL (Agunbiade et al. 2012).

3.4 *ALLIUM CEPA* LINN. *(AC)*

Common names: Bulb Onion, Common Onion (Figure 3.5)

A. cepa, a member of the garlic family, is also known as the bulb onion or common onion, a vegetable that is the most widely cultivated species of the genus *Allium*. It has long been used as dietary supplements for the traditional treatment of diabetes in Asia, Europe, and the Middle East (Bailey and Day 1989). Allyl propyl disulphide (APDS) (Figure 3.6a) and diallyl disulphide (Figure 3.6b) have been identified in the onion to be the active components in hyperglycaemic attenuation (Dey, Attele

FIGURE 3.5 Illustration of *Allium cepa*.

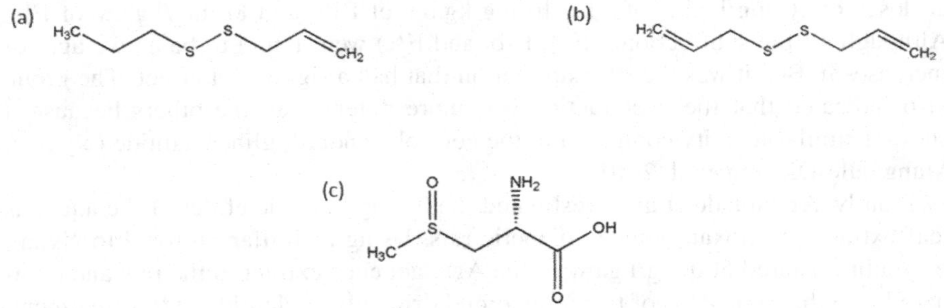

FIGURE 3.6 Structural formulae of isolated compounds from *Allium cepa*: (a) allyl propyl disulphide, (b) diallyl disulphide, and (c) S-methyl cysteine sulphoxide.

and Yuan 2002); however, their excessive intake has been known to have detrimental effects on hepatic metabolism (Augusti and Benaim 1975, Jain and Vyas 1974). Researchers have postulated that APDS functions as a competitive inhibitor to insulin for insulin-inactivating sites in the liver, thereby increasing the concentration of free insulin in the body.

Other studies using concentrated extracts from the plant organs have reported weak hypoglycaemic effects in healthy and alloxan-induced diabetic animals and healthy human subjects (Augusti and Benaim 1975, Jain and Vyas 1974, Jain, Vyas and Mahatma 1973, Day and Bailey 1986, Mathew and Augusti 1975).

Kumari et al. reported another isolate (S-methyl cysteine sulphoxide [SMCS] [Figure 3.6c]) that also showed antidiabetic activity. The administration of SMCS isolate at 200 mg/kg.bw for 45 days to "alloxanized" rats significantly controlled BG and lipids in serum and tissues and normalized the activities of liver hexokinase, glucose 6-phosphatase, and 3-hydroxy-3-methyl-glutaryl-coenzyme A (HMG-CoA) reductase. The effect of this hyperglycaemic attenuation was notably comparable to that of glibenclamide and insulin (Kumari, Mathew and Augusti 1995).

Finally, a clinical study by Sharma et al. showed that the oral administration of aqueous onion extracts (at 25, 50, 100, and 200 g) to overnight-fasted healthy volunteers (30 minutes before, after, or simultaneously giving 50 g of an oral glucose supplement) significantly increased glucose tolerance in a dose-dependent manner. The effect was noted to be comparable to that of their control standard drug, tolbutamide (K. K. Sharma et al. 1977).

3.5 *ALLIUM SATIVUM* LINN. *(AS)*

Common names: Garlic, Lahasun (Figure 3.7)

A. sativum is a species in the onion genus, *Allium*. Its close relatives include the onion, shallot, leek, chive, and Chinese onion. Native to Central Asia and northeastern Iran, it has long been used as a common seasoning worldwide, with a history of several thousand years of human consumption and use.

Many studies have been undertaken to evaluate the hyperglycaemic attenuating capability of AS extracts via oral administration. Jain et al. reported that the oral administration of its organic extract (at 0.25 g/kg.bw) caused a maximum reduction

FIGURE 3.7 Illustration of *Allium sativum*.

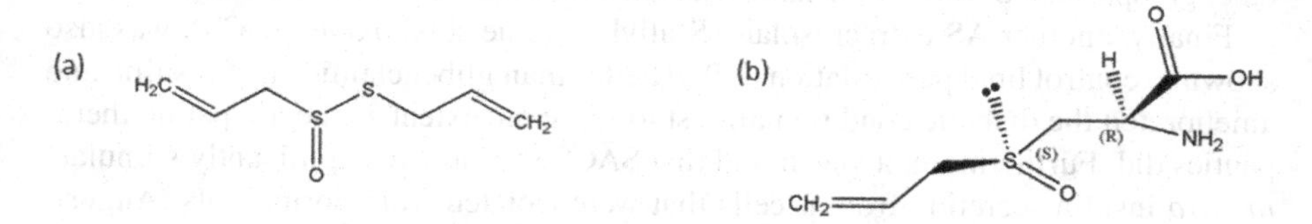

FIGURE 3.8 Structural formulae of isolated compounds from *Allium sativum*: (a) allicin and (b) alliin.

in BG levels by 26.2%, in alloxan-induced diabetic rabbits (Jain and Vyas 1975). In another study, the oral administration of allicin (isolated from AS) (Figure 3.8a) at 0.25 g/kg.bw exhibited hypoglycaemic effects that were comparable to tolbutamide in mildly diabetic rabbits (BG ranging from 180 to 300 mg%); however, there was no such effect in severely diabetic animals (BG >350 mg%) (Mathew and Augusti 1973).

The Zacharias group later demonstrated that the oral administration of an aqueous homogenate of garlic (10 mL/kg.bw/day) to sucrose-fed rabbits had significant ameliorative effects on FBG, protein levels, and triglyceride levels in serum, liver, and aorta, while increasing hepatic glycogen (HG) and free amino acid contents (Zacharias et al. 1980).

Other studies have also shown that the oral administration of 1 mL/kg.bw garlic juice extract was able to effect a 68% reduction in elevated PG levels in diabetic rats (El-Gamal and El-Khedr 2017). With similar findings being reported by El-Demerdash et al., there is further support for the inclusion of raw garlic in a regular diet for the purpose of BG regulation (El-Demerdash, Yousef and El-Naga 2005).

The Kasuga group reported that the oral pretreatment of stress-induced hyperglycaemic mice with aged garlic extracts (AGE) (5 and 10 mL/kg.bw) prevented adrenal hypertrophy and hyperglycaemia and the elevation of cortisone without altering serum insulin levels. Interestingly, the efficacy of AGE was the same as that of diazepam (5 mg/kg.bw). This supported that AGE may prevent the stress-induced risk of DM and its progression. Furthermore, daily oral feeding of garlic extract (100 mg/kg.bw) increased the cardiovascular functions in STZ rats, prevented abnormality in the lipid profile, and increased fibrinolytic activities with decreased platelet aggregation. Notably, there was an increase in plasma insulin levels with a concomitant decrease in PG levels. Of equal interest, the group reported that, daily oral feeding of the same dose for 16 weeks showed anti-atherosclerotic effects in their STZ diabetic rats (Patumraj, Tewit et al. 2000).

Sheela et al. reported that the administration of alliin (an analogue to allicin) (Figure 3.8b) at a dosage of 200 mg/kg.bw significantly decreased the concentration of serum lipids and BG, and the activities of serum enzymes such as alkaline phosphatase, acid phosphatase, lactate dehydrogenase, and liver glucose-6-phosphatase. It also significantly increased liver and intestinal HMG-CoA reductase activity and liver hexokinase activity (Sheela and Augusti 1992). In a later study, the oral administration of alliin to alloxan-induced diabetic rats for 1 month showed similarity in improvements to their glucose intolerance, weight loss, and the depletion of liver glycogen, when compared to other groups treated with glibenclamide and insulin (Sheela, Kumud and Augusti 1995).

Finally, another AS extract isolate, S-allyl cysteine sulphoxide (SACS), was also shown to control lipid peroxidation (LPO) better than glibenclamide and insulin, and ameliorated the diabetic condition almost to the same extent as the allopathic therapeutics did. Furthermore, it was found that SACS was able to significantly stimulate *in vitro* insulin secretion from β-cells that were isolated from normal rats (Augusti and Sheela 1996).

3.6 *ALOE BARBADENSIS MILL.* (ABM)

Common names: Aloe, Aloe vera, Alovis, Sabila (Figure 3.9)

A. barbadensis is a short-stemmed succulent perennial herb of the Liliaceae family that can grow to a height of 3–4 m through its central axis or stem. Different species can be identified by variations in their leaf size and flower colour (I. A. Ross 2003, Pamplona-Roger 2005). Native to North Africa, the Mediterranean region of southern Europe, and the Canary Islands, it is now cultivated throughout tropical regions including the West Indies (I. A. Ross 2003).

For centuries, many aloe species have been used to treat several conditions due to their laxative, anti-inflammatory, immunostimulant, antiseptic wound- and burn-healing, antiulcer, antitumour, and antidiabetic activities (Okyar et al. 2001). ABM is claimed to contain several polysaccharides, whose biological interactions result in hypoglycaemic effects (Bailey and Day 1989, Ghannam et al. 1986). Several researchers have isolated some of these compounds (β-sitosterol Figure 3.10a], campesterol [Figure 3.10b] (Verma et al. 2018), lophenol [4-methylcholest-7-en-3-ol] [Figure

FIGURE 3.9 Illustration of *Aloe barbadensis.*

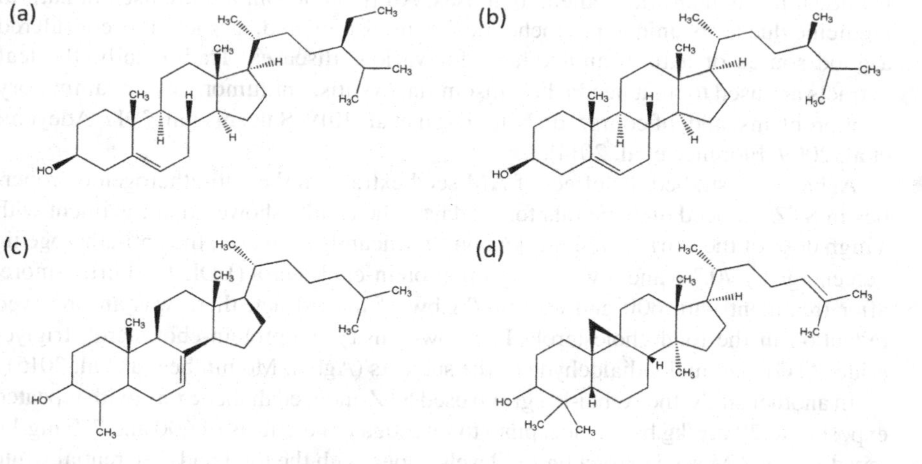

FIGURE 3.10 Structural formulae of isolated compounds from *Aloe barbadensis Miller*: (a) β-sitosterol, (b) campesterol, (c) lophenol (4-methylcholest-7-en-3-ol), and (d) cycloartenol (9,19-cyclolanostan-3-ol).

3.10c], and cycloartenol [9,19-cyclolanostan-3-ol] [Figure 3.10d] [Shakib et al. 2019, Pothuraju et al. 2016], 24-methyl-lophenol, 24-ethyl-lophenol, and 24-methylene-cycloartanol [Pothuraju et al. 2016]) from the ABM plant, and tested for their relative activities and toxicities.

Two independent groups led by Okyar and Ghannam, respectively, investigated the impact of leaf pulp and leaf gel extracts of ABM on non-diabetic, T1DM, and T2DM rats. While both groups reported confirmed hypoglycaemic activity in their diabetic subjects, using the pulp extracts, the Okyar group in particular, observed an enhanced effect for the ABM-treated T2DM subjects, in comparison to their glibenclamide-treated controls. Interestingly, it was found that ABM was ineffective in lowering the BG level in non-diabetic rats, contrary to the effect of glibenclamide (Okyar et al. 2001, Ghannam et al. 1986).

Similarly, a study conducted by Rajasekaran et al. demonstrated the hypoglycaemic activity of alcoholic extracts of ABM gel. The extract was found to maintain glucose homeostasis by controlling the activity of the carbohydrate metabolizing enzymes. It was observed that the effect of the gel extract in this study was similar to that of Okyar et al., where the effect of the ABM gel extract on the BG levels of non-diabetic rats was not deemed significant (Shakib et al. 2019, Rajasekaran et al. 2004). It was further proposed that the potentiation of insulin from the β-cells of the islets of Langerhans, or its release from its bound form, may be the possible mechanism by which ABM brings about its hypoglycaemic action (Rajasekaran et al. 2004).

3.7 *ANNONA MURICATA* LINN. *(AM)*

Common names: Soursop, Prickly Custard Apple, Graviola, Guanabana (Figure 3.11)

A. muricata is a tropical plant species that belongs to the family: Annonaceae, and is known for its many ethnomedicinal uses. All parts of *Annona* are used in natural medicine due to its unique phytochemical composition and, as such, it is considered a good source of natural antioxidants for various diseases. Traditionally, the leaf extracts are used to treat headaches, insomnia, cystitis, antitumor, anti-inflammatory liver problems, and, of course, diabetes (Agu et al. 2019, Saleem et al. 2017, Adeyemi et al. 2009, Florence et al. 2014).

Agbai et al. studied the effect of AM seed extract on the anti-atherogenic properties in STZ-induced diabetic rats for 30 days. The results showed that treatment with a high dose of the extract (750 mg/kg.bw) significantly increased the anti-atherogenic percentage by 70.7% and low-density lipoprotein-cholesterol (LDL-C). Furthermore, after treatment with 600 and 800 mg/kg.bw of the extract, there was an observed reduction in the total cholesterol (TC), low-density lipoprotein-cholesterol, triglycerides (TG), and malondialdehyde of the subjects (Agbai, Mounmbegna et al. 2015).

In another study, the Agbai-led group used STZ-induced diabetic rats (with repeated exposure to 20 mg/kg.bw of clozapine) to investigate the effects of 600 and 750 mg/kg .bw doses of AM seed extract on BG levels, along with the total and differential white cell count. From their analysis of the data, the group concluded that the seed extract

FIGURE 3.11 Illustration of *Annona muricata*.

successfully exerted a hypoglycaemic effect on the clozapine-treated rats, without improving the decreased total and white cell count (Agbai, Njoku et al. 2015).

Adewole et al. then looked at the effects of the aqueous leaf extract on the morphology of pancreatic β-cells and oxidative stress induced in STZ- induced diabetic rats. The findings of the study not only indicated that AM treatment had ameliorative effects on pancreatic tissues, which were subjected to STZ-induced oxidative stress, but also showed the protection and preservation of pancreatic β-cell integrity (Adewole and Ojewole 2009).

Alternatively, studies have revealed that FOXO1 protein inactivation in the nucleus is one of the targets in the treatment of DM. Considering this, through an *in silico* study, Dini et al. aimed at predicting the inhibition of the FOXO1 protein by AM's active compounds. Their results showed that the compounds contained in AM: xylopine (Figure 3.12a), anonaine (Figure 3.12b), isolaureline (Figure 3.12c),

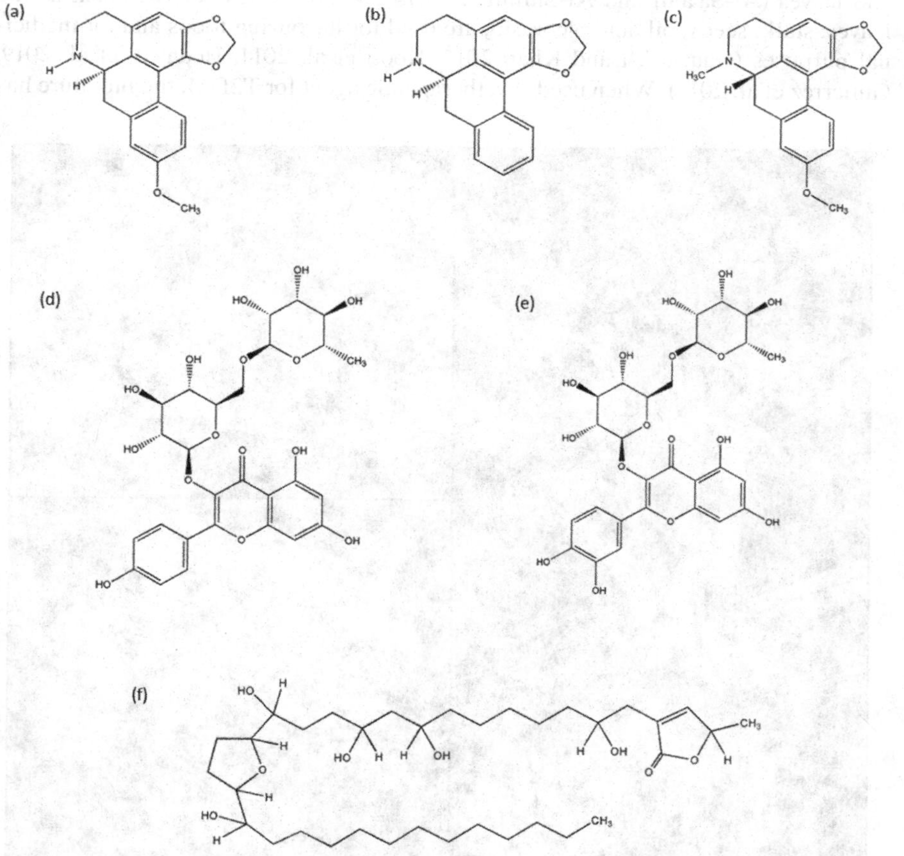

FIGURE 3.12 Structural formulae of isolated compounds from *Annona muricata*: (a) xylopine, (b) anonaine, (c) isolaureline, (d) kaempferol 3-O-rutinoside, (e) rutin, and (f) muricatocin.

kaempferol 3-O-rutinoside (Figure 3.12d), rutin (Figure 3.12e), and muricatocin (Figure 3.12f) had the ability to strongly and spontaneously bind with the active side of the FOXO1 protein, and consequently inhibit its activity (Damayanti, Utomo and Kusuma 2017).

3.8 *APIUM GRAVEOLENS* LINN. *(AG)*

Common names: Ajmod, Celeriac, Celery, Leaf celery, Root celery, Wild celery (Figure 3.13)

A. graveolens is an erect, annual or perennial herb of the Apiaceae or Umbelliferae family (Gauri, Ali and Khan 2015). It is usually located in Europe, temperate and subtropical parts of Africa and Asia (Gauri, Ali and Khan 2015, AlMalaak, Almansour and Hussein 2018, Kooti et al. 2014), and North America (Kooti et al. 2014). The whole plant is known for its specific taste and aromatic smell, especially the roots and leaves (Al-Sa'aidi and Al-Shihmani 2013). Various forms of AG, such as the leaves, stalk, seeds, oil, and oleoresin, are used for flavouring foods and for medicinal purposes (Gauri, Ali and Khan 2015, Kooti et al. 2014, Hedayati et al. 2019, Gutierrez et al. 2014). When used as a therapeutic agent for T2DM, the literature has

FIGURE 3.13 Illustration of *Apium graveolens*.

shown AG to stimulate the pancreatic secretion of insulin and, as such, reduce BG levels (Kooti et al. 2014).

Mans and Aburjai investigated the hypoglycaemic effects of AG on diabetic rats. They found that the AG seed extract (at a dose of 425 mg/kg.bw) led to a significant decrease in BG levels and also influenced the increase in serum insulin levels. It was equally interesting to note that for the duration of the experiment, the group reported that the rats displayed minimal toxicity – associated symptoms (Mans and Aburjai 2019).

Our research into the bioactive components of the AG plant, which are responsible for its ameliorative effects in DM therapy, revealed the compounds: luteolin (Figure 3.14a), apigenin (Figure 3.14b) (Hedayati et al. 2019, Niaz, Gull and Zia 2013, Yusni et al. 2018), and 3-*n*-butylphthalide (Figure 3.14c) (Hedayati et al. 2019). We found that the outcomes from the application of such therapy included a reversal of the catabolic features of insulin deficiency, a decrease in the release of glucagon, or an increase in the release of insulin (Al-Sa'aidi and Al-Shihmani 2013, Mans and Aburjai 2019, Niaz, Gull and Zia 2013). Furthermore, reports have been made of the direct stimulation of glycolysis in peripheral tissues, an increase in the removal of BG (Mans and Aburjai 2019, Niaz, Gull and Zia 2013), or a reduction in the absorption of glucose from the gastrointestinal tract (Mans and Aburjai 2019).

As reported by Al-Sa'aidi et al., the "drenching" of STZ-induced diabetic rats with a 60 mg/kg.bw dose of n-butanol AG seed extract resulted in a reduction in the BG levels and also influenced the regeneration of damaged pancreatic islets (Al-Sa'aidi and Al-Shihmani 2013). In support of this, another study revealed that AG seed extract caused a significant decrease in serum glucose levels and the

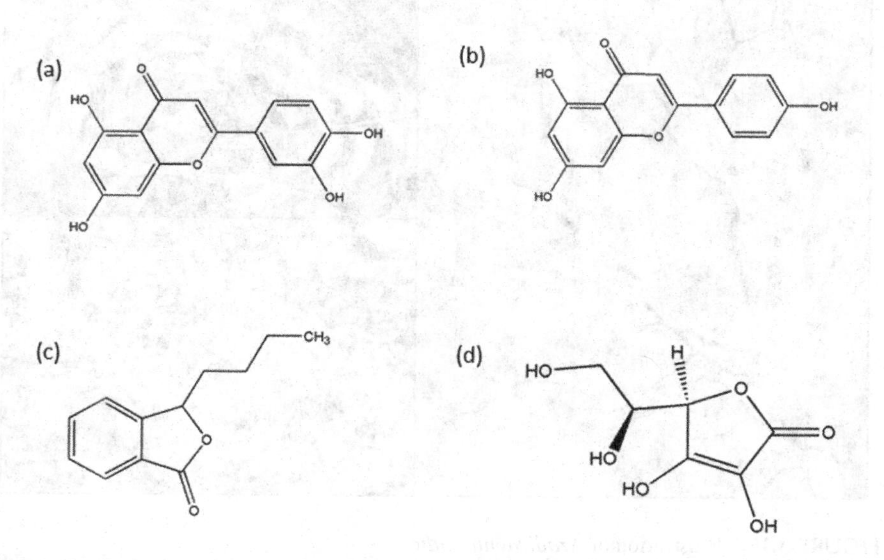

FIGURE 3.14 Structural formulae of isolated compounds from *Apium graveolens*: (a) luteolin, (b) apigenin, (c) 3-*n*-butylphthalide, and (d) vitamin C.

induction of insulin release from the pancreatic islets (Hedayati et al. 2019, Niaz, Gull and Zia 2013). However, Wasfi and AL-kabi claimed that the effect of the AG extract may take a long time to show (possibly due to its low metabolism) and suggested that the extract lowers BG levels by affecting the absorption of glucose in the intestine and not by stimulating the production of insulin by the pancreas (Wasfi and AL-kabi 2019).

Apart from the seeds, the leaf extracts from the AG plant also exhibit hypoglycaemic activity (Gutierrez et al. 2014, Abozid, El-Rahman and Mohamed 2018). Stemming from their study, Abozid et al. proposed that the glucose-lowering effect of AG leaf extracts can be attributed to the presence of a high content mixture of phenolic compounds, vitamin C (Figure 3.14d), and flavonoids (Figure 3.4), all of which have been associated with antidiabetic activity in other botanical sources (Abozid, El-Rahman and Mohamed 2018).

3.9 *AZADIRACHTA INDICA* LINN. *(AI)*

Common names: Dogoyaro, Indian lilac, Margosa tree, Neem, Nimtree (Figure 3.15)

A. indica, popularly known as Neem, is a member of the mahogany family: Meliaceae. It is a fast-growing tropical evergreen tree that grows to 20 m in height. It was originally cultivated in the Asian subcontinents, and later introduced to the Caribbean and South and Central America. Different parts of the tree (such as the

FIGURE 3.15 Illustration of *Azadirachta indica.*

leaves, bark, and flowers) have been studied for its use in phytomedicinal therapy. Stemming from these studies, supportive evidence has come forward for its therapeutic use in the treatment of diabetes, infertility, bacterial infection, skin disease, oral care, and scalp disorders. The literature has also revealed the nature and efficacy of its active compound: azadirachitin (Figure 3.16), and has shown that alongside other active constituents, has been able to confer the antidiabetic activity of the plant (Pandey 2020).

Shravan et al. showed that a single dose of AI aqueous leaf extract (250 mg/kg.bw) was able to reduce elevated BG levels in diabetic rats after 24 hours, and this effect was observed to continue until the end of the 14-day study period (Dholi et al. 2011).

In another study, Bhat et al. compared the effects of chloroform extracts of AI against those of aqueous/methanolic extracts of *Bougainvillea spectabilis* Linn. (a flowering plant), using STZ-induced diabetic mice. At the end of the 21-day study, the group reported a significant increase in glucose-6-phosphate dehydrogenase activity as well as hepatic and skeletal muscle glycogen content. In their immunohistochemical analysis, they observed that treatment with both extracts resulted in the regeneration of the insulin-producing cells and there was a corresponding increase in the plasma insulin and *c*-peptide levels. From their analysis, it was concluded that both sources were good candidates for the development of new nutraceutical treatments for diabetes (Bhat et al. 2011).

Finally, Satyanarayana et al. administered dehydrated aqueous leaf-extracts (400 mg/kg.bw) to STZ-induced diabetic rats for 30 days. The group reported that the animals receiving the therapeutic dose of the extract showed a normalization in the altered levels of BG, serum insulin, lipid profile, and insulin signalling molecules, as well as GLUT-4 proteins (Satyanarayana et al. 2015).

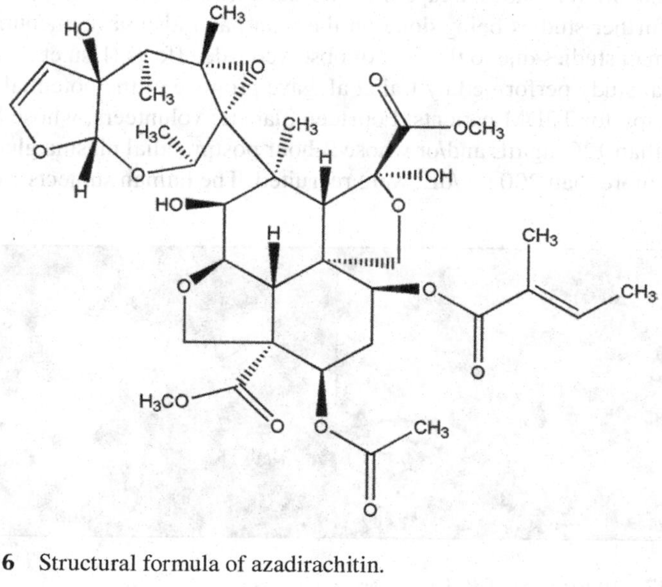

FIGURE 3.16 Structural formula of azadirachitin.

3.10 *BIDENS PILOSA* LINN. *(BiP)*

Common names: Black Jack, Cobbler's pegs, Farmers friend, Hairy beggarticks, Needle bush, Spanish needle, Sticky beaks (Figure 3.17)

B. pilosa is an erect, perennial herb with serrated, lobed, or in a dissected form of green opposite leaves, white or yellow flowers, and long, narrow, ribbed, black seeds. It is believed to have originated in South America, before spreading to other parts of the world. Regarding its pharmacological actions, its antidiabetic polyynes have been reported to be an effective treatment for both T1DM and T2DM. Yang et al. used the "bioactivity-directed isolation and identification" approach to identify three active polyynes: 3-β -D-glucopyranosyl-1-hydroxy-6(E)-tridecene-8,10,12-triyne (Figure 3.18a), 2-β -D-glucopyranosyloxy-1-hydroxy-5(E)-tridecene-7,9,11-triyne (Figure 3.18b), and 2-β -D-glucopyranosyloxy-1-hydroxytrideca-5,7,9,11-tetrayne (cytopiloyne) (Figure 3.18c), as well as three index compounds: 4,5-di-O-caffeoylquinic acid (Figure 3.18d), 3,5-di-O-caffeoylquinic acid (Figure 3.18e), and 3,4-di-O-caffeoylquinic acid (Figure 3.18f), from the butanolic extract (BE) of the BP plant (W.-C. Yang 2014).

An animal study by Hsu et al. assessed the hypoglycaemic activity of aqueous whole plant extract (BiPWE) using a population of C57BL/KsJ-db/db mice. An evaluation of the attenuating effects was determined through analysis of the BG levels, serum insulin, HbA1c, glucose tolerance, and islet structure. The results showed that after the 28-day study period, treatment with BiPWE (50–250 mg/kg.bw) significantly improved glucose tolerance, decreased HbA1c levels, and protected the islet structure in the diabetic mice. The investigating team proposed that the extract's mechanism of action was similar to that of the control standard, glimepiride. However, unlike the prescription drug, which is incapable of either restoring or protecting the pancreatic cells, the BiPWE showed capability for such action. The findings of the group supported further studies being done on the plant, and also gave encouragement to attempt human studies due to the lack of observed side effects (Hsu et al. 2009).

A clinical study performed by Lai et al. gave promise of the potential use of BiP extract therapy for T2DM patients. Fourteen diabetic volunteers, whose FBG levels were more than 126 mg/dL and/or whose 2-hour postprandial plasma glucose (PPG) levels were more than 200 mg/dL, were recruited. The human subjects were divided

FIGURE 3.17 Illustration of *Bidens pilosa*.

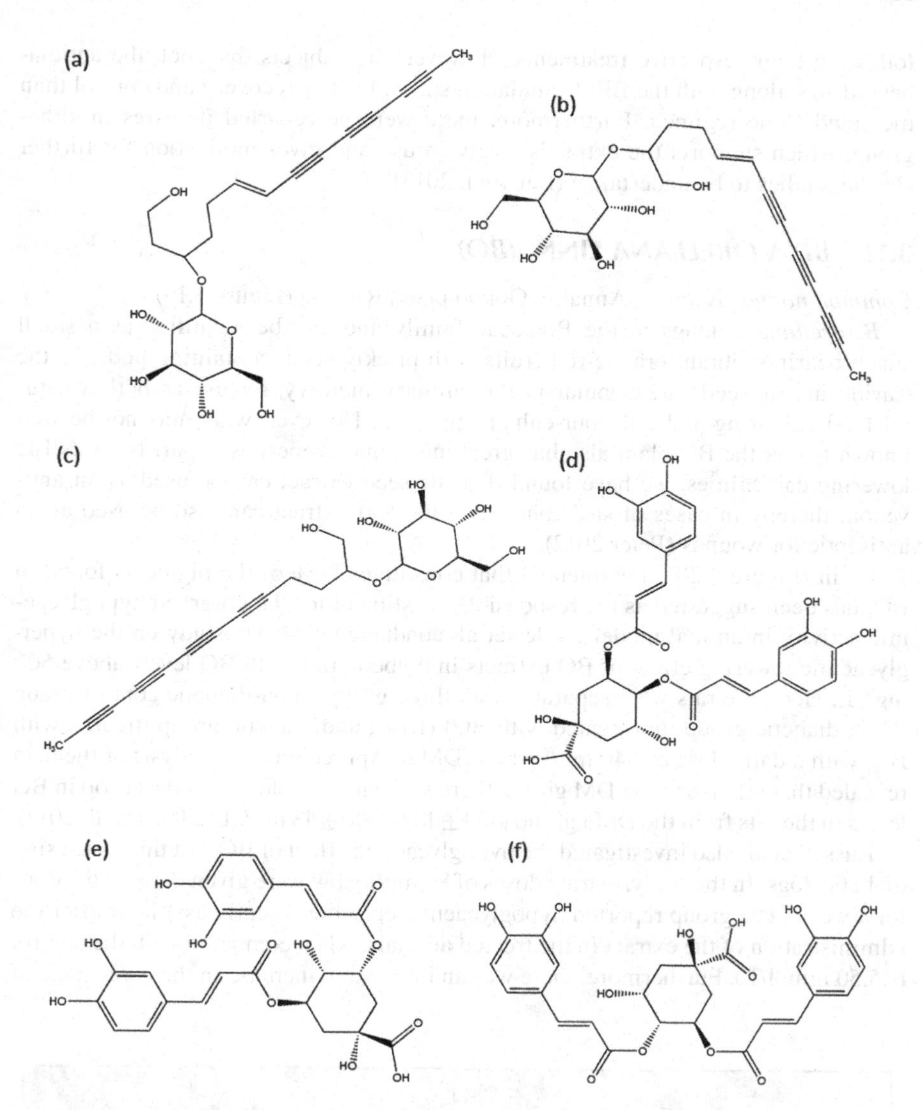

FIGURE 3.18 Structural formulae of isolated compounds from *Bidens pilosa*: (a) 3-β-D-glucopyranosyl-1-hydroxy-6(E)-tridecene-8,10,12-triyne, (b) 2-β-D-glucopyranosyloxy-1-hydroxy-5(E)-tridecene-7,9,11-triyne, (c) 2-β-D-glucopyranosyloxy-1-hydroxytrideca-5,7,9,11-tetrayne, (d) 4,5-di-O-caffeoylquinic acid, (e) 3,5-di-O-caffeoylquinic acid, and (f) 3,4-di-O-caffeoylquinic acid.

into two groups: Group 1 consisted of six diabetics, who were prescribed 400 mg oral doses of the BiP formulation for 3–7 months, while Group 2 comprised the remaining eight diabetics, who were allowed to continue their regular antidiabetic drug regimen in addition to prescribed doses of the BiP formulation. It was found that both groups displayed noticeable decreases in their hyperglycaemic levels by following their respective treatments. However, the subjects that took the antidiabetic drugs along with the BiP formulation showed better recovery and control than the stand-alone regimen. Furthermore, there were no reported illnesses in either group, which supports the extract's safety in use and gives motivation for further similar studies to be undertaken (Lai et al. 2015)

3.11 *BIXA ORELLANA* LINN. *(BO)*

Common names: Achiote, Annatto, Ookoo plant, Roucou (Figure 3.19)

B. orellana belongs to the Bixaceae family and can be identified as a small shrub bearing vibrant orange-red fruits with prickly seed–containing pods. In the Caribbean, its seeds are popular in the culinary industry, for use as both a natural food colouring and a flavour-enhancing agent. However, what may not be well known is that the BO plant also has great medicinal properties. Apart from its BG lowering capabilities, we have found that the seed extract can be used as an antivenom therapy in cases of snakebites, and the leaf extract can also be used as an antiseptic for wounds (Peter 2012).

Bixin (Figure 3.20), a carotenoid that constitutes 80% of the pigments found in BO, has been suggested as the responsible constituent for the observed hypoglycaemic activity in animal models. Teles et al. conducted a 14-day study on the hyperglycaemic lowering effect of BO extracts in diabetic rats with BG levels above 500 mg/dL. Here, the rats were separated into three groups: non-diabetic control group (CN); diabetic group (not treated with BO) (DM); and diabetic group (treated with BO, with a daily dose of 540 mg/kg.bw) (DMa). Appreciatively, analysis of the data revealed that relative to the DM group, there was a noticeable but modest drop in BG levels in the rats from the DMa group (423 ± 18 to 380 ± 43 mg/dL) (Teles et al. 2014).

Russell et al. also investigated the hypoglycaemic effect of BO, but this time using diabetic dogs. In the study, extract doses of 80 mg/kg.bw were given daily to the dogs for a week. The group reported hypoglycaemic episodes as early as 1 hour after the administration of the extract in the treated animals, when compared with the control (<5.50 mmol/L). Furthermore, there was an interesting increase in the percentage of

FIGURE 3.19 Illustration of *Bixa orellana*.

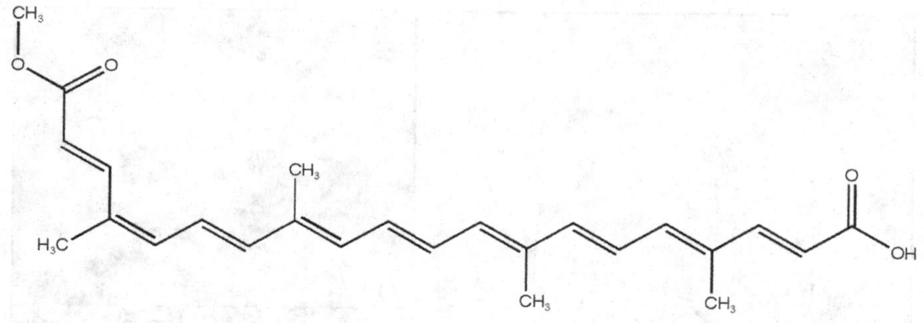

FIGURE 3.20 Structural formula of bixin.

insulin levels of the extract-treated dogs (>55 μIU/mL) when compared to those of the control subjects (Russell, Morrison and Ragoobirsingh 2005).

The results from both studies show supporting evidence for the antidiabetic activity of BO extracts. However, more studies are encouraged to confirm the potency and possible associated toxicity, if the plant is to be used as a viable form of T2DM therapy.

3.12 *BRASSICA JUNCEA* LINN. *(BJ)*

Common names: Brown mustard, Mustard greens, Chinese mustard, Leaf mustard, Rai (Figure 3.21)

B. juncea is a member of the Cruciferae family and is native to parts of Asia and the Middle East. However, it is now also cultivated worldwide in regions such as America, Russia, and parts of Africa (Kumar et al. 2011). Its leaves, roots, and shoots are known as vegetables, while its seeds were initially used as a condiment or a source of oil (Szollosi 2011). It is interesting to note that the climatic zones in which the plant has now spread to, have been seen to influence the phytochemical content of the seeds. This, in turn, has extended its use as a source of natural medicine in various cultures. Although there is information suggesting that BJ-derived products have therapeutic potential against diabetes, no definitive statements on the nature of the phytoconstituents involved can yet be made (Kumar et al. 2011).

One of the active constituents of the BJ seed extract has been suggested to be sulphoraphane (Figure 3.22a), which is thought to increase the activity of glycogen synthetase (Verma et al. 2018). That aside, Yokozawa et al. also investigated the effect of isorhamnetin 3,7-diglucoside (Figure 3.22b) (the main compound in BJ), which has also been confirmed to exhibit antidiabetic activity (Yokozawa et al. 2002).

Other studies have been carried out, aimed at demonstrating the BG-lowering activity of the AG aqueous seed extracts, using alloxan-induced diabetic rats with doses of 250, 350, and 450 mg/kg.bw extract, over both short-term and long-term periods (Verma et al. 2018, Kumar et al. 2011, Thirumalai et al. 2011). It was

FIGURE 3.21 Illustration of *Brassica juncea*.

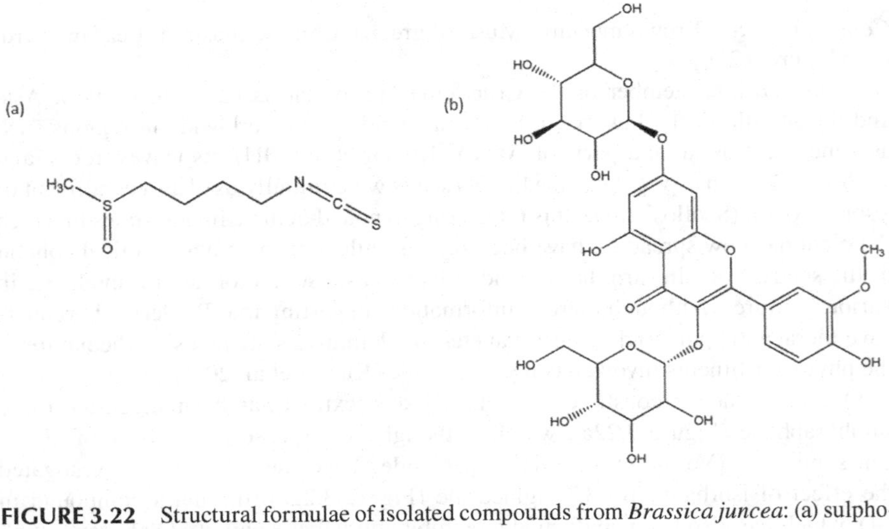

FIGURE 3.22 Structural formulae of isolated compounds from *Brassica juncea*: (a) sulphoraphane and (b) isorhamnetin 3,7-diglucoside.

observed that the insulin serum augmenting effect was dose dependent, with the highest therapeutic dose being 450 mg/kg.bw (Thirumalai et al. 2011).

Additionally, the results of the study carried out by Yadav et al. confirmed that BJ can play a role in the management of the prediabetic state of insulin resistance and its use in higher quantities as a food ingredient should be encouraged for those prone to diabetes (S. P. Yadav et al. 2004).

In summary, although extracts or parts of BJ are not recommended as a stand-alone therapy for the control of severe diabetes and may not be beneficial in insulin-dependent diabetes, it can be used as a safe dietary supplement in the management of a prediabetic or a moderately diabetic state. Its safety in use can be supported by the fact that it has been consumed over centuries by people without effect and, more recently, efficacy studies using rodent subjects have reported no adverse effects on food intake and various haematological parameters (Grover, Yadav and Vats 2002).

3.13 *BRYOPHYLLUM PINNATUM* LINN. *(BP)*

Common names: Good Luck, Resurrection plant, Wonder of the World (Figure 3.23)

B. pinnate is a crassulaceant herb of about 1 metre in height, with opposing glabrous leaves (each having three to five deeply crenulated, fleshy leaflets). Its use as a form of herbal remedy for an array of human disorders, including hypertension, diabetes mellitus, bruises, wounds, boils, abscesses, insect bites, arthritis, rheumatism, joint pains, headaches, and body pains, is ubiquitous throughout the world. Its range of therapeutic applications can be justified by its large number of active compounds (such as flavonoids [Figure 3.4], glycosides, steroids, bufadienolides, and organic acids).

A study carried out by Aransiola et al., using a population of 25 diabetic rats, demonstrated the glucose-lowering ability of its ethanolic extract. In an effort to highlight the extract's potency, the activity was gauged against that of glibenclamide. The animals were divided into five groups and made diabetic by injecting them with Glucose-*D* at a dose of 3 g/kg.bw. Groups 2–4 were then administered varying doses of the extract, with Group 5 being administered the glibenclamide drug and Group 1 remaining as the untreated control. At the end of the experiment, the results showed good correlation in the activities of the extracts, with the glibenclamide standard; the highest performance observed was at the dosage of 200 mg/kg.bw (Aransiola et al. 2014).

FIGURE 3.23 Illustration of *Bryophyllum pinnate*.

In another study by Ojewole, the glucose-lowering effect of the aqueous extracts of BP leaves was weighted against that of another commercial antidiabetic drug, chlorpropamide. Using a population of 24 young, adult Wistar rats, the animals were randomly divided into two groups (Group A: test and Group B: control). Diabetes was induced in the Group A rodents by intraperitoneal injections of STZ, while Group B rats were treated with a placebo (distilled water). The Group A subjects were then orally treated with either varying doses of the BP extract (ranging from 25 to 800 mg/kg.bw) or chlorpropamide (250 mg/kg.bw). Arising from this, Ojewole reported that when compared with Group B, the pretreated rats of Group A showed significant reductions in their BG levels. The hypoglycaemic effect of the plant extract was significant at 1–2 hours following oral administration, reaching the peak of its hypoglycaemic effect at 2–4 hours after administration, and remaining stable for a further 8 hours (Ojewole 2005).

3.14 *CAPPARIS SPINOSA* LINN. *(CaS)*

Common names: Alaf-e-Mar, Alcaparro, Caper bush, Cappero, Kebbar, Wild Watermelon (Figure 3.24)

C. spinosa is a deep-rooted, spiny, perennial shrub that can grow up to 10 m in height. It is native to, and widely distributed throughout the Middle East, but is now cultivated worldwide in tropical and subtropical zones (Nabavi et al. 2016, Zhang and Ma 2018, Vahid, Rakhshandeh and Ghorbani 2017).

The wide range of pharmacological applications of CaS can be justified by its well-known history as an antifungal, antileishmanial, antihepatotoxic, anti-inflammatory, and antidiabetic agent (Eddouks, Lemhadri and Michel 2004). Regarding the latter, studies have shown that the root, seeds, and aerial components contain compounds that exhibit hypoglycaemic effects (Zhang and Ma 2018, Vahid, Rakhshandeh and Ghorbani 2017, Okur et al. 2018). As a matter of fact, the pickled forms of the CaS fruits and flower buds are consumed by diabetic patients in the Middle East because of the belief that it possesses BG-lowering and hypolipidaemic properties (Huseini et al. 2013).

FIGURE 3.24 Illustration of *Capparis spinosa*.

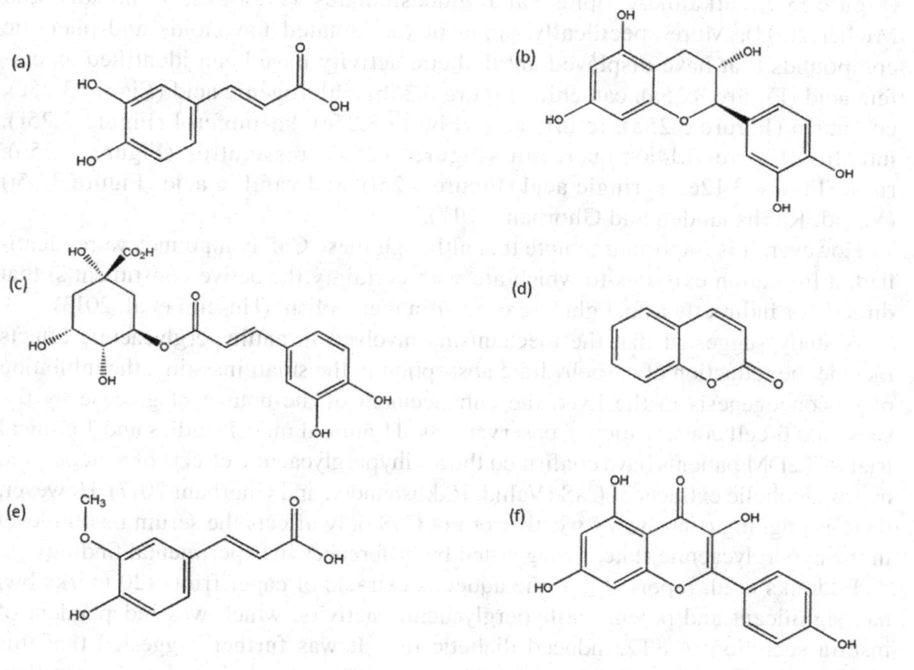

FIGURE 3.25 Structural formulae of isolated compounds from *Capparis spinosa*: (a) caffeic acid, (b) catechin, (c) chlorogenic acid, (d) coumarin, (e) ferulic acid, (f) kaempferol, (g) quercetin, (h) resveratrol, (i) syringic acid, and (j) vanillic acid.

The main constituents of CaS have been demonstrated to be flavonoids (Figure 3.4), alkaloids, lipids, and glucosinolates (Eddouks, Lemhadri and Michel 2004). More specifically, some of the isolated flavonoids and phenolic compounds that have displayed antidiabetic activity have been identified as caffeic acid (Figure 3.25a), catechin (Figure 3.25b), chlorogenic acid (Figure 3.25c), coumarin (Figure 3.25d), ferulic acid (Figure 3.25e), kaempferol (Figure 3.25f), luteolin (Figure 3.14a), quercetin (Figure 3.25g), resveratrol (Figure 3.25h), rutin (Figure 3.12e), syringic acid (Figure 3.25i), and vanillic acid (Figure 3.25j) (Vahid, Rakhshandeh and Ghorbani 2017).

However, it is important to note that although these CaS compounds were identified, a limitation exists as to which are with certainty the active constituent(s) that directly or indirectly affect glucose or insulin metabolism (Huseini et al. 2013).

A study suggested that the mechanisms involved in antihyperglycaemic effects include the reduction of carbohydrate absorption in the small intestine, the inhibition of gluconeogenesis in the liver, the enhancement of the uptake of glucose by tissues, and β-cell conservation. Conservatively, 11 animal model studies and 1 clinical trial in T2DM patients have confirmed the antihyperglycaemic effects of aqueous and hydro-alcoholic extracts of CaS (Vahid, Rakhshandeh and Ghorbani 2017). However, there is ongoing controversy whether or not CaS only affects the serum insulin level in the hyperglycaemic state, as suggested by differences in experimental findings.

Eddouks et al. reported that the aqueous extracts of caper fruits (20 mg/kg.bw) had significant and potent antihyperglycaemic activity, which was independent of insulin secretion in STZ-induced diabetic rats. It was further suggested that this activity was due to hepatic glucose production and/or stimulation of glucose utilization by peripheral tissues, particularly muscle and adipose tissue (Zhang and Ma 2018, Eddouks, Lemhadri and Michel 2004).

Currently, there is limited information on the adverse effects of CaS on the human body. However, independent toxicity studies performed by Zhang et al. and Huseini et al. reported no signs of hepatotoxicity, nephrotoxicity, or any other adverse effects (Zhang and Ma 2018, Huseini et al. 2013).

3.15 *CARICA PAPAYA* LINN. *(CaP)*

Common names: Aanabahe-hindi, Buah papaya, Papaw, Papaya, Pawpaw (Figure 3.26)

C. papaya is a member of the Caricaceae family. It is popularly known as pawpaw in the West Indies, and is one of the most popular tropical fruits, cultivated in tropical and subtropical areas across the world. It is a small, fast-growing evergreen tree, that can grow as high as 10 m. The tree's leaves, fruits, stem, and latex have been used for a variety of medicinal therapies for ailments including diabetes, gastrointestinal disorders, and wound healing (as an anti-inflammatory or antimicrobial agent) (I. A. Ross 2003). Although there are few *in vivo* clinical trials aimed at investigating the antidiabetic effects of this plant, it is known that the edible fruit contains high levels of vitamin A (Figure 3.27a), vitamin C (Figure 3.14d), and the phytochemical compound β-cryptoxanthin (Figure 3.27b), which can regulate BG levels through diet (Venkateshwarlu et al. 2013).

FIGURE 3.26 Illustration of *Carica papaya*.

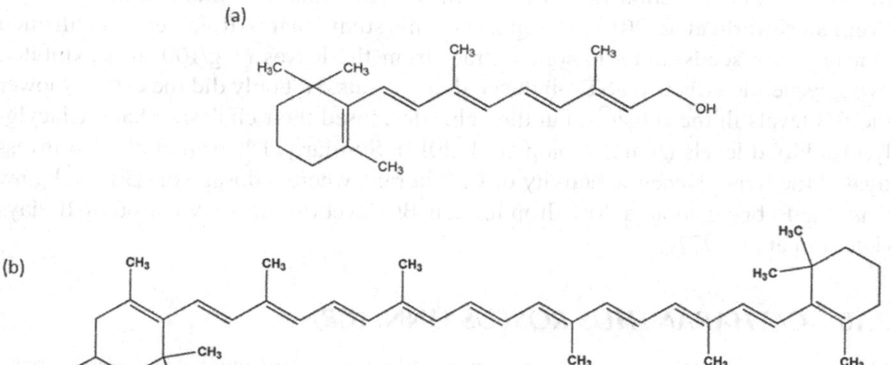

FIGURE 3.27 Structural formulae of isolated compounds from *Carica papaya*: (a) vitamin A and (b) β-cryptoxanthin.

Papaya is one of the many fruits ranked low on the glycaemic index and, as such, diabetics can consume it safely in small portions, thus reducing the risk of complications associated with the disease (Ismawanti, Suparyatmo and Wiboworini 2019). As a matter of fact, the results from a clinical study showed that the consumption of as much as 438 g of the unripened fruit maintained normoglycaemic levels in a test group of diabetic patients (Fatema et al. 2011).

In addition to the previously mentioned components, flavonoid compounds (Figure 3.4) have also been isolated in papaya. These flavonoids are able to control glucose levels by regenerating pancreatic β-cells, which in turn increases the production of insulin in the body (Akhlaghi and Foshati 2017).

Furthermore, flavonoids also inhibit the GLUT-2 transporter (in the intestinal mucosa) and the α-glucosidase enzyme (in the small intestine). They have also been reported to reduce glucose absorption and prevent carbohydrate breakdown, which would ultimately result in reduced BG levels (Ismawanti, Suparyatmo and Wiboworini 2019).

Maniyar and Bhixavatimath experimented with CaP leaf extracts on alloxan-induced diabetic rats and found that, at a dose of 400 mg/kg.bw, the aqueous extract was effective in reducing the BG levels of the diabetic animals (Maniyar and Bhixavatimath 2012). Similarly, Airaodion et al. also investigated the antidiabetic activity of the compounds contained in papaya leaves. In their study, they gauged the efficacy of the compounds against that of the commercial drug, glibenclamide. The group reported that due to the comparable activity with the commercial drug, the use of papaya leaf extracts could be a viable alternative in the treatment of diabetes (Airaodion et al. 2019).

To complete our story, we report on the interesting findings of Venkateshwarlu et al., who directed their investigations towards the aqueous extract of the CaP seeds. The group found that a therapeutic dose of 200 mg/kg.bw of the extract resulted in a significant reduction in BG levels in STZ/nicotinamide-induced diabetic rats (Venkateshwarlu et al. 2013). In support of this study, Juárez-Rojop et al. confirmed that both the seeds and ethanolic extract from the leaves (3 g/100 mL) exhibited hypoglycaemic effects on STZ-induced diabetic rats. Not only did the extracts lower the BG levels in the subjects, but they also decreased their cholesterol and triacylglycerol blood levels (Juárez-Rojop et al. 2012). Similarly, Dhawan et al. also investigated the hypoglycaemic activity of CaP in rats, where a dosage of 250 mg/kg.bw was able to bring about a 30% drop in their BG level over a study period of 10 days (Dhawan et al. 1977).

3.16 *CATHARANTHUS ROSEUS* LINN. *(CR)*

Common names: Bright eyes, Cape periwinkle, Graveyard plant, Madagascar periwinkle, Old maid, Pink periwinkle, Rose periwinkle (Figure 3.28)

C. roseus, also known as *Vinca rosea* Linn., is an ornamental sub-shrub that is cultivated mainly for its alkaloids: vinblastine (Figure 3.29a), vincristine (Figure 3.29b), and ajmalicine (Figure 3.29c). Reports have shown that these isolated compounds display both anticancer and hypotensive activity; however, there are records of it also being used for DM therapy (Jaleel et al. 2006, Council of Scientific & Industrial Research 1992, Kulkarni et al. 1999). Furthermore, the CR plant is known to produce more than 100 monoterpenoids and indole alkaloids, which claim to have ameliorative effects in different organs (Jordan, Thrower and Wilson 1991).

Although Watt and Breyer-Brandwijk considered that most investigations into the claim of hypoglycaemic activity have shown negative results and that any advantage obtained can be ascribed to weak digitalis and purgative action (Watt and Breyer-Brandwijk 1932), the plant still enjoys a widespread reputation

FIGURE 3.28 Illustration of *Catharanthus roseus*.

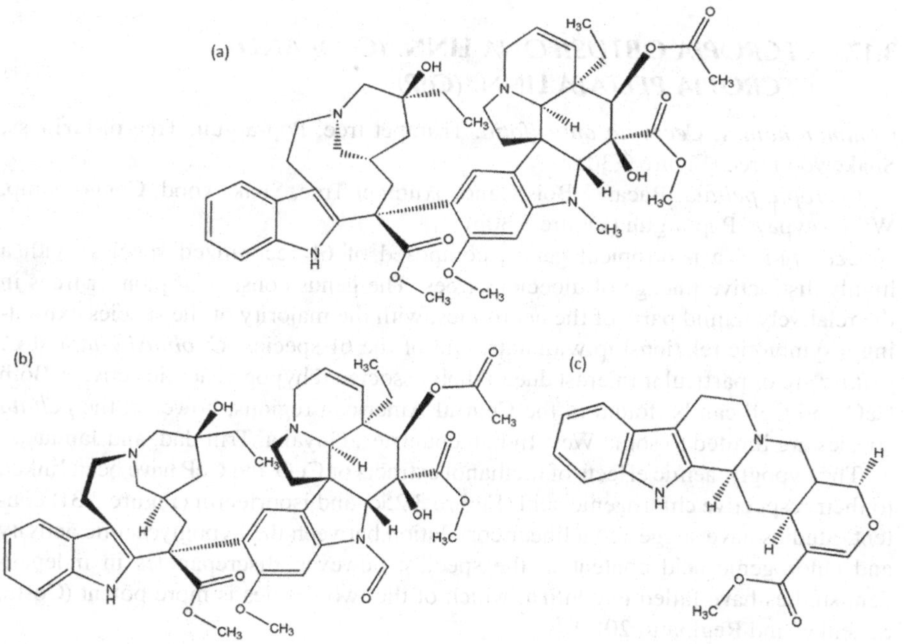

FIGURE 3.29 Structural formulae of isolated compounds from *Catharanthus roseus*: (a) vinblastine, (b) vincristine, and (c) ajmalicine.

for its antihyperglycaemic action in the treatment of diabetes (Watt and Breyer-Brandwijk 1932, Pillay, Nair and Santi Kumari 1959, Nammi et al. 2003, Singh et al. 2001). It is proposed that the mechanism of action of CR is mediated either through the enhancement of insulin secretion from the β-cells of Langerhans or through an extra-pancreatic mechanism (Rahimi 2015, Nammi et al. 2003). Using a population of healthy rats in their first study, Chattopadhyay et al. reported that the oral administration of varying dosages of water-soluble fractions of the ethanolic extract of CR leaves (100, 250, 500, and 1000 mg/kg.bw) showed a significant dose-dependent reduction in BG at 4 hours by 26.22%, 31.39%, 35.57%, and 33.37%, respectively. Furthermore, the oral administration of CR extract (at 500 mg/kg.bw), 3.5 hours before an oral glucose tolerance test (OGTT) (10 g/kg.bw) and 72 hours after STZ administration (50 mg/kg.bw) in rats showed a significant antihyperglycaemic effect (Chattopadhyay et al. 1991).

In yet another study it was reported that the oral administration of the leaf and twig extract (500 mg/kg.bw) was beneficial to animals by lowering their elevated BG levels (Malvi et al. 2011). Finally, histopathological studies carried out by Ahmed et al. showed the effective reversal of the alloxan-induced changes in the BG level and the β-cell population in the pancreas. These findings support previous studies and encourage more research on the therapeutic benefits of CR in DM therapy (Ahmed et al. 2010).

3.17 *CECROPIA OBTUSIFOLIA* LINN. *(CeO) AND CECROPIA PELTATA* LINN. *(CeP)*

Common names: *Cecropia obtusifolia*: Trumpet tree, Pop-a-gun, Tree-of-laziness, Snakewood tree (Figure 3.30a);

Cecropia peltata: Bacano, Bois Cano, Trumpet Tree, Snakewood, Congo pump, Wild pawpaw, Pop-a-gun (Figure 3.30b)

Cecropia is a neotropical genus, composed of 61 recognized species, with a highly distinctive lineage of dioecious trees. The genus consists of pioneer trees in the relatively humid parts of the neotropics, with the majority of the species exhibiting a symbiotic relationship with ants. Out of the 61 species, *C. obtusifolia* and *C. peltata* are of particular interest due to their associated hypoglycaemic activity. Both CeO and CeP can be found in the Central American regions; however, the *peltata* species are limited to some West Indian countries, Guyana, Trinidad, and Jamaica.

The hypoglycaemic effects of methanol extracts of CeO and CeP have been linked to their respective chlorogenic acid (Figure 3.25c) and isoorientin (Figure 3.31) content. Studies have suggested a linear correlation between the hypoglycaemic activity and chlorogenic acid content in the species; however, discrepancies in independent studies have failed to confirm which of the two species is more potent (Costa, Schenkel and Reginatto 2011).

The mechanism of the antidiabetic activity of the genus was investigated by evaluating the inhibitory effect of the butanolic extract of CeO on α-glycosidase. *In vitro* studies confirmed the inhibitory activity of the extract on α-glycosidase, while the results from the *in vivo* experiment showed that the extract was able to reduce the PG levels as early as 90 minutes after administration. Another investigation into the mechanism of action, demonstrated an increase in *in vitro* glucose reuptake (in both insulin-sensitive and insulin-resistant adipocytes), when treated with an aqueous extract of CeO, and with chlorogenic acid (Costa, Schenkel and Reginatto 2011). Recently, it was reported that the oral administration of both aqueous and

FIGURE 3.30 Illustrations of *Cecropia* species: (a) *Cecropia obtusifolia* and (b) *Cecropia peltata.*

FIGURE 3.31 Structural formula of isoorientin.

butanolic extracts of CeO and CeP was able to reduce hepatic glucose production *in vivo* and inhibit the activity of glucose-6-phosphatase during the process of gluconeogenesis. The study brought to light the attractive BG-lowering efficiency of the *Cecropia* extracts in relation to the commercial drug, metformin (Andrade-Cetto and Cárdenas-Vázquez 2010).

Studies on human patients are described only for CeO. The daily intake of aqueous leaf extracts by 22 patients (with a history of poor responses to conventional allopathic treatments) resulted in a significant reduction in glucose levels. In another study that lasted 32 weeks, 12 patients who were diagnosed with T2DM, showed a significant and sustained hypoglycaemic effect over a treatment period of 18 weeks; this effect was maintained for a further 4 weeks after the treatment ended (Costa, Schenkel and Reginatto 2011).

Finally, having established the hypoglycaemic activity of the *Cecropia* extracts, the Andrade-Cetto group set out to gauge the relative potencies of both aqueous and butanolic extracts of CeP against that of glibenclamide. The group was able to report a good correlation in activity between the commercial drug and both extracts, with the greater potency observed for the butanolic extracts versus the aqueous type (Andrade-Cetto, Cárdenas and Ramírez-Reyes 2007).

3.18 *CENTELLA ASIATICA* LINN. *(CeA)*

Common name: Gotu Kola, Asiatic Pennywort, Spade leaf (Figure 3.32)

With its origins traced back to Asia, *C. asiatica* can now be found flourishing in swampy areas and ditches of the Caribbean. Research on the plant has led to the isolation of a large variety of biomolecules, which include flavonoids (Figure 3.4), tannins, phytosterols, saponins, amino acids, and sugars, each of which commands various ethnopharmacological applications inclusive of DM therapy (Dewi and Maryani 2015).

In an attempt to elucidate the active compounds and the mechanism of action responsible for the antidiabetic activity in CeA, Dewi et al. tailored an *in vitro* experiment that investigated the suspected α-glucosidase inhibiting action of its extracts.

FIGURE 3.32 Illustration of *Centella asiatica*.

The group was successful in isolating two compounds from the organic extracts: kaempferol (Figure 3.25f) and quercetin (Figure 3.25g).

Furthermore, based on observed increased insulin levels and the results from pancreatic histopathology studies of diabetic models, the antidiabetic activity of those samples was believed to occur by stimulating the non-damaged pancreatic β-cells by another isolate: Asiaticoside (Figure 3.33), to produce more insulin (Fitrianda et al. 2017).

Another group aimed to examine the effect of dosage variations of whole plant extracts on STZ/nicotinamide-induced diabetic rats. The results confirmed the antidiabetic effect of CeA extract dosages of 1200 mg/kg.bw/day and reported a similarity in activity with that of the control standard, metformin (at 45 mg/kg.bw/day) (Muhlishoh, Wasita and Nuhriawangsa 2018).

Finally, in a study using alloxan-induced diabetic male Wistar rats, Emran et al. confirmed both the glucose and total cholesterol-lowering ability of CeA extracts at a dosage of 50 mg/kg.bw. Equally interesting was their observation that the plant extracts showed no toxic effects on the treated rats. This finding suggested that

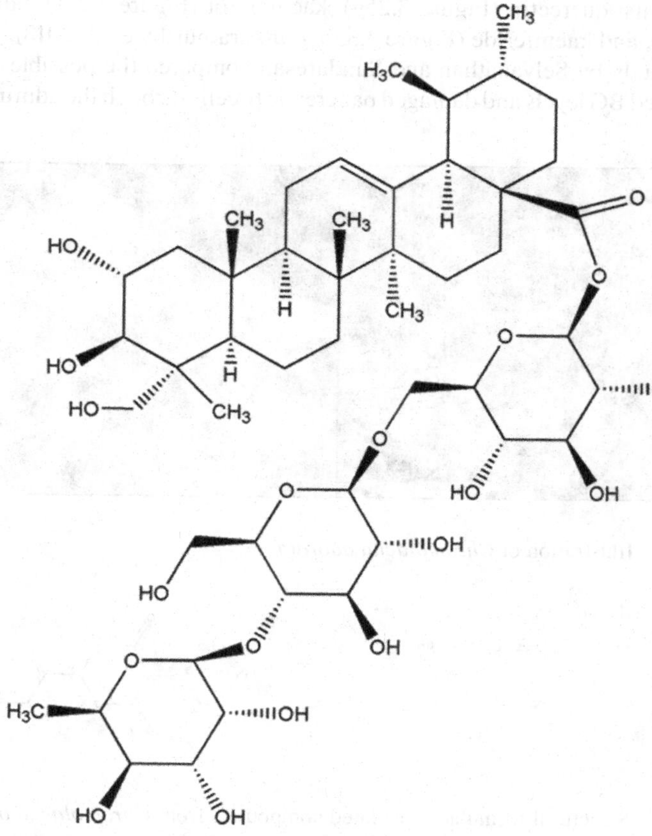

FIGURE 3.33 Structural formula of asiaticoside.

unlike insulin and other common antidiabetic agents, overdose of the drug may not result in hypoglycaemia (Emran et al. 2015).

3.19 *CHROMOLAENA ODORATA* LINN. (CO)

Common names: Siam weed, Christmas bush, Jack in the Box, Devil Weed, Communist Pacha, Common Floss Flower, Rompe Saragüey (Figure 3.34)

C. odorata (also classified as *Eupatorium odoratum* Linn.) is a flowering shrub of the Astereaceae family that is considered one of the world's worst weeds (Omonije, Saidu and Muhammad 2019, Oguntibeju 2019, Amaliah et al. 2019). It is a fast-growing, densely tangled bush, with a height range of 2–10 m (Oguntibeju 2019, Ijioma et al. 2014, Ogugua et al. 2013). The CO plant can be found thriving in all parts of the United States, Mexico, West Africa, the Caribbean (Oguntibeju 2019), and other tropical countries (Asomugha et al. 2013).

Many studies in the past, focused on the antidiabetic properties of CO extracts. Justifiably, we have found that greater attention was placed on the CO leaves than its roots (Omonije, Saidu and Muhammad 2019) due to the presence of the bioactive components: quercetin (Figure 3.25g), kaempferol (Figure 3.25f), tamarixetin (Figure 3.35a), and kaemferide (Figure 3.35b) (Onkaramurthy et al. 2013).

The first study by Selvanathan and Sundaresan compared the possible improvement of elevated BG levels and damaged pancreatic β-cells through the administration

FIGURE 3.34 Illustration of *Chromolaena odorata*.

FIGURE 3.35 Structural formulae of isolated compounds from *Chromolaena odorata*: (a) tamarixetin and (b) kaempferide.

of ethanolic leaf extracts to diabetic albino rats. The results showed that on the ninth day of treatment, doses of 200 and 400 mg/kg.bw led to lowered serum glucose levels (from 258.2 ± 8.5 and 259.9 ± 10.9 to 204.9 ± 9.2 and 191.2 ± 8.4 mg/dL, respectively) (Selvanathan and Sundaresan 2020). Another investigating team reported that infused CO leaf extract caused a reduction in the BG levels of male mice, with the most effective dosage being at a concentration of 20% (Amaliah et al. 2019).

In a final study, Omonije et al. examined the BG levels of diabetic rats when treated with methanolic root extracts of CO. The group reported a BG level reduction of 49.86% and 68.3% at doses of 300 and 600 mg/kg.bw, respectively, thus again supporting the potential therapeutic use of CO in DM therapy (Omonije, Saidu and Muhammad 2019).

It should be noted that CO contains carcinogenic pyrrolizidine alkaloids (Ogugua et al. 2013). Although acute toxicity studies revealed that CO methanolic root extracts and ethanolic leaf extracts did not produce significant changes in animal behaviour in doses up to 5000 mg/kg.bw (Omonije, Saidu and Muhammad 2019, Ijioma et al. 2014), the finding was not supported by Asomugha et al., who observed a 43% mortality rate in their test rats after oral administration of the aqueous leaf extract, at a dose of 1077 mg/kg.bw (Asomugha et al. 2013). The conflicting reports clearly underscores the need for further toxicity and safety-in-use studies for this plant.

3.20 *CITRUS AURANTIIFOLIA* LINN. *(CiA)*

Common names: Lime, Key Lime, Mexican Lime, Mexican Thornless Key Lime (Figure 3.36)

FIGURE 3.36 Illustration of *Citrus aurantiifolia*.

C. aurantiifolia is a short, densely branched evergreen tree that is popular for its edible fruit. It is mainly grown in tropical and subtropical regions, such as the Caribbean, Mexico, Florida, and Southeast Asia. Belonging to the Rutaceae family, the fruit is known for its vibrant green and yellow exterior (peel) and the zesty, acidic taste of its juice. Like its cousin, the lemon (CiL), CiA shares a wide array of applications in both the culinary and medical fields. Regarding the latter, the leaves, fruit, peel, and essential oils are used because of their confirmed astringent, antibacterial, and anti-inflammatory properties (Fern n.d.). While it is common knowledge that this citrus family is packed full of powerful antioxidants, such as vitamin C (Figure 3.14d), much research has been carried out to confirm the effectiveness of the other constituent chemicals as a natural diabetic treatment.

One study conducted by Mawatri et al. focused on evaluating the synergistic effects of mixtures of CiA extracts with that of *Cinnamomum burmannii* Linn. (CB; cinnamon) to treat hyperglycaemic levels in diabetic rats. Twenty-five STZ-induced rats were divided into five groups and administered different doses of the extracts for 30 days: Group 1: negative control, standard diet; Group 2: positive control, high-cholesterol diet; Group 3: high-cholesterol diet + CiA extract (100 mg/kg.bw/day) and CB extract (200 mg/kg.bw/day); Group 4: high-cholesterol diet + CiA extract (300 mg/kg.bw/day) and CB extract (400 mg/kg.bw/day); and Group 5: high-cholesterol diet + CiA extract (500 mg/kg.bw/day) and CB extract (800 mg/kg.bw/day). The results showed that both extracts had a significant effect on the BG in Groups 2, 3, and 5, decreasing their levels by 53.2% overall. Also, the levels of superoxide dismutase (SOD) increased in Group 4 indicating higher anti-inflammatory agents in the body. They concluded that, when used in conjunction with CB extracts, CiA can be an effective therapy in the management of diabetic complications in hyperglycaemic patients (Mawarti, Khotimah and Rajin 2018).

In another study, Karmi and Nasab compared the antidiabetic effect of varying concentrations of CiA juice against that of a combination of garlic extract (AS) and CiA juice. Gauging their results against metformin (as a control standard), it was found that a 50% dilution of the CiA juice was able to comparably, but temporarily, reduce the BG levels in diabetic rats. Furthermore, they reported that the combination of CiA juice (50%) and garlic (250 mg/100 g.bw) also significantly lowered the BG levels in the diabetic subjects (Karimi and Nasab 2014).

Ibrahim et al. experimented with the essential oil from the CiA leaf on alloxan-induced diabetic rats for 14 days. They conducted a chemical analysis of the oil and identified the main constituent as *D*-limonene (Figure 3.37) (57.84%), which according to the literature, has confirmed anti-inflammatory and antimicrobial properties.

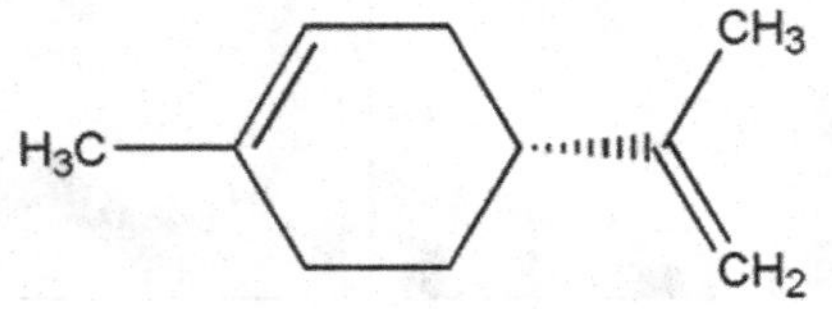

FIGURE 3.37 Structural formula of *D*-limonene.

The essential oil was hydro-distilled and administered to the rats daily, at doses of 25, 50, 100, and 200 mg/kg.bw. From this, the group reported that the oil (at 100 mg/kg.bw) showed a significant reduction in and normalization of the FBG and hepatic glucose during the 14-day study, as compared to the other doses (Ibrahim et al. 2019).

3.21 *CITRUS LIMON* LINN. *(CiL)*

Common name: Lemon (Figure 3.38)

C. limon belongs to the Rutaceae family alongside lime, grapefruit, and orange. It is grown in tropical and semi-tropical climates and is recognized as an oval citrus fruit with either a smooth porous or rough skin. Its high content of ascorbic acid (vitamin C), potassium, and citric acid makes them very popular in the West for treating the common cold and sore throat, detoxifying the body, and weight loss. The leaves, stems, juices, pulp, and peel from the fruit contain vitamins that have anti-inflammatory and antioxidant properties. Furthermore, the high content of polyphenols present in the fruit and the peel of CiL confers an appreciable hypoglycaemic effect (Rafique et al. 2020).

Lv et al. investigated the antihyperglycaemic activity of CiL using the hydro-alcoholic extract found in its peel. Using a population of STZ-induced diabetic rats, the subjects were divided into four groups: normal control, diabetic control, diabetic + low dose of the peel extract, and diabetic + high dose of the peel extract. The rats were then administered 250–500 mg/kg.bw of the extract daily for 35 days. After the 35-day period, both groups that were treated with the low and high dosage of the peel extract, showed a significant decrease in FBG levels and a reduction in

FIGURE 3.38 Illustration of *Citrus limon*.

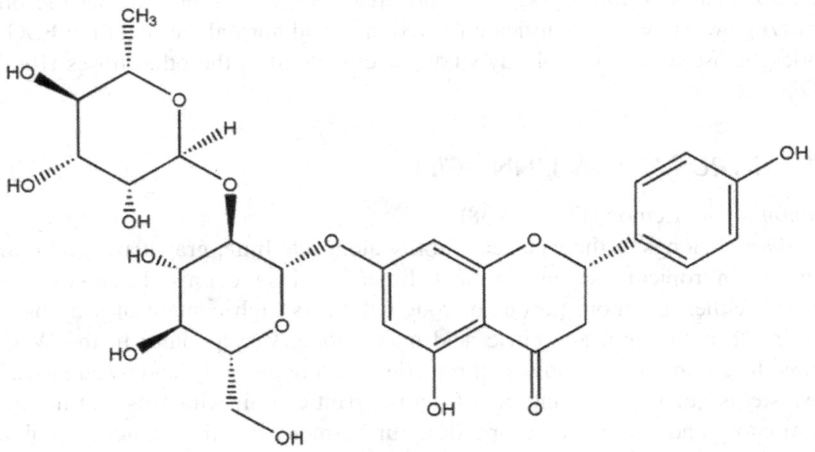

FIGURE 3.39 Structural formula of naringin.

insulin resistance. Interestingly, there was an increase in body weight in the treated diabetic rats, while oxidative stress was alleviated and antioxidant activities in the liver were restored (Lv et al. 2018).

Another group demonstrated the ameliorative effects of CiL and pomegranate juices on the BG and plasma insulin levels in alloxan-induced diabetic rats. The rodents were administered three doses of CiL juice: 0.2, 0.4, and 0.6 mL/kg.bw over a 6-week period, after which a significant decrease in the BG levels was noted. It was proposed that the antihyperglycaemic effect was due to the high content of vitamin C (Figure 3.14d), flavonoids (Figure 3.4), and naringin (Figure 3.39) (Riaz, Khan and Ahmed 2013).

Naim et al. also investigated the antidiabetic activity of CiL peel on alloxan-induced diabetic rats using a hexane extract and compared its effect against the commercial drug, glimepiride. The rats were separated into three groups: Group 1 (treated with normal saline), Group 2 (treated with 15 μg/kg.bw glimepiride), and Group 3 (treated with 10 mg/kg.bw hexane extract from CiL peel). After 1 week, the results indicated that the hexane extract reduced the BG levels by 44.57%, 75.96%, 95.43%, and 98.08% in 24, 48, 72, and 96 hours, respectively. The investigating team found that the efficacy of the extract showed good correlation with that of glimepiride and was able to confirm its therapeutic potential in DM treatment (Naim et al. 2012).

3.22 *CITRUS PARADISI* LINN. *(CiP)*

Common names: Grapefruit, Paradise Citrus (Figure 3.40)

The citrus fruit *C. paradisi* was developed in the West Indies in the early 1700s as a cross-breed between an orange and a shaddock. It is consumed worldwide, not only because of its taste and nutritional value, but also because of its accredited medicinal

FIGURE 3.40 Illustration of *Citrus paradisi.*

properties (Owira and Ojewole 2009). Evergreen grapefruit trees have an average height of 5–6 m, but can grow as tall as 13–15 m. The tree has long, thin, glossy, dark-green leaves, and produces white, four-petaled flowers that are approximately 5 cm wide. The spheroidal fruit is usually yellow skinned, with an average diameter of 10–15 cm (Wikipedia 2021), and the fruit pulp is either yellow, pink, or white in colour.

Since the 1930s, CiP has been a constituent of many diets (as an anti-obesity ingredient). Such an inclusion therefore suggests that consumption of grapefruit or grapefruit juice may be beneficial to T2DM patients. Modern research has confirmed its effectiveness in influencing an improvement in glucose tolerance as well as plasma lipid (PL), TC, LDL, and TG levels (Hayanga et al. 2015). Drawing from these studies, it would be expected that the fruit and its extracts would top the list of herbal therapeutants for the disease. However, research has uncovered that the consumption of CiP juice is connected to unfavourable increases in the plasma concentrations of orally administered drugs, due to the inhibition of the activity of the metabolizing enzyme, cytochrome P450IIIA4 (CYP3A4) (Owira and Ojewole 2009, Lundahl et al. 1997, Saito et al. 2005, Zaidenstein et al. 1998). As such, although only a few cases of life-threatening effects have been reported, the use of CiP juice as a viable T2DM therapeutant has since been avoided (Hayanga et al. 2015).

The components in CiP juice that are responsible for CYP3A4 inhibition have not been fully determined. However, *in vitro* studies have confirmed the CYP3A4 inhibitory action of some CiP juice isolates, such as bergamottin (Figure 3.41a) and 6′,7′-dihydroxy-bergamottin (Figure 3.41b) (Christensen et al. 2002).

Despite all the aforementioned negativities, researchers continue to explore the potential of its use in DM therapy. A study by Hayanga et al. demonstrated an

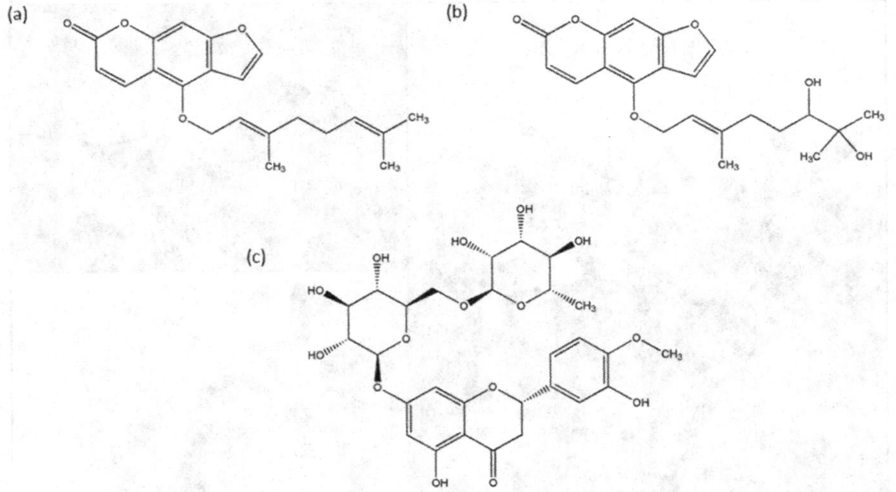

FIGURE 3.41 Structural formulae of isolated compounds from *Citrus paradisi*: (a) bergamottin, (b) 6′,7′-dihydroxy-bergamottin, and (c) hesperidin.

improvement in the glucose intolerance of STZ-induced diabetic rats, using CiP juice therapy. The group reported an improvement in the weight of the treated rats as well as more normalized FBG and fasting plasma insulin (FPI) levels, when compared to the control animals. This observation suggested that the juice therapy may have inhibited protein and lipid catabolism, which is associated with insulin deficiency. Furthermore, the untreated diabetic rats showed significantly reduced hepatic glycogen content compared to the non-diabetic controls. Interestingly, treatment with either the CiP juice alone or in combination with insulin significantly increased the hepatic glycogen content in relation to the untreated group. However, it was observed that the juice had no effect on pancreatic insulin secretion, which confirmed that the bioactive constituents of CiP are neither insulinotropic nor did they promote pancreatic β-cell regeneration. Finally, the group noted that the effects of CiP juice appeared to be enhanced by insulin. As such, it was speculated that the juice, like metformin, could be upregulating adenosine monophosphate–activated protein kinase (AMPK), which is known to activate glucokinase and simultaneously deactivate G6Pase and PEPCK, respectively (refer to the actions of biguanides in Chapter 2) (Hayanga et al. 2015).

In another study, Jung et al. investigated the effect of the contained bioflavonoids (hesperidin [Figure 3.41c] and naringin [Figure 3.39]) on the BG levels, hepatic glucose-regulating (HGR) enzyme activity, hepatic glycogen concentration, and plasma insulin levels of genetically raised diabetic mice (C57BL/KsJ-db/db). Test animals were fed an AIN-76 control diet (standard diet formulation proposed by the American Institute of Nutrition) supplemented with hesperidin (0.2 g/kg.bw diet) or naringin (0.2 g/kg.bw diet). It was found that the treated animals showed a significant reduction in BG levels, but elevated HG activity and HG concentrations, when compared with the control group. The results of the experiment gave good support to both hesperidin's and naringin's important roles in preventing the progression of hyperglycaemia, partly by increasing hepatic glycolysis and glycogen concentration and/or by lowering hepatic gluconeogenesis (Jung et al. 2004).

While the ameliorative effects of CiP juice seem to be promising for T2DM models, its juice contents have been proven ineffective in T1DM subjects. The metformin-like mechanism of action, which was proposed by Hayanga, can be corroborated with the findings of Xulu's group, which confirmed that although naringin showed signs of anti-atherogenic activity, it was ineffective in antihyperglycaemic action (Xulu and Owira 2012).

3.23 *CITRUS SINENSIS* LINN. *(CiS)*

Common names: Orange, Sweet orange (Figure 3.42)

Along with lemon, *C. sinensis* belongs to the Rutaceae family. CiS is a small, evergreen, aromatic tree that is grown in both tropical and subtropical regions. It is more round in shape compared to CiL and comes in various colours and sizes. It is popular for its fruit's juices and essential oils, which are good sources of vitamin C and antioxidants. These components carry powerful medicinal properties that can be used in the treatment of the common cold. However, besides the common cold, studies have shown that the extracts from the plant (juices, peels, and bark) exhibit antihyperglycaemic activity, and can be applied to T2DM therapy.

Parmar and Kar investigated the antidiabetic effect of the peel extract in alloxan-induced diabetic mice over a period of 10 days. Four different concentrations of the extract, ranging between 12.5, 25, 50, and 100 mg/kg.bw, were administered daily to

FIGURE 3.42 Illustration of *Citrus sinensis.*

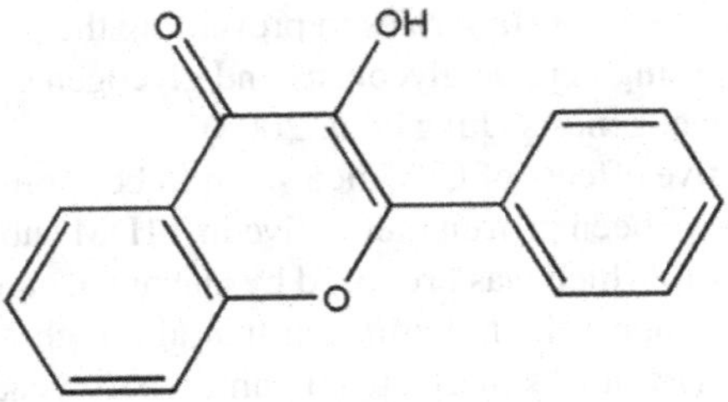

FIGURE 3.43 General structural formula of flavonols.

the subjects, which resulted in a dose responsive arrest of the hepatic lipid peroxidation, as well as a significant decrease in serum glucose (Parmar and Kar 2007). From this, the study group suggested that the medley of compounds present in the CiS extract, such as flavonols (Figure 3.43), hesperidin (Figure 3.41c), and naringin (Figure 3.39), were responsible for its antidiabetic and antiperoxidative activity (Parmar and Kar 2007, 2008).

Next, Sathiyabama et al. investigated the antidiabetic effect of the methanolic fruit peel extract in STZ insulin-resistant diabetic rats over a 30-day period. The rodents were divided into five groups: Group 1: normal control rats; Group 2: diabetic control; Group 3: diabetic (treated with 50 mg/kg.bw of CiS methanol extract); Group 4: diabetic (treated with 100 mg/kg.bw of CiS methanol extract); and Group 5: diabetic (treated with 100 mg/kg.bw of metformin). The group reported that the diabetic rats (receiving both 50 and 100 mg/kg.bw of methanol extract) showed decreased BG levels at 60 and 120 minutes after glucose administration. Additionally, the peroxisome proliferator-activated receptor gamma (PPARγ), GLUT-4, and insulin receptor messenger ribonucleic acid (IR-mRNA) expression levels in CiS methanol extract–treated diabetic groups were significantly higher than those in the diabetic control group. These results thus influenced the team's proposal that the observed activity was due to the presence of polyphenols, which were abundant in the methanolic elixir (Sathiyabama et al. 2018).

In an independent study, Boris et al. experimented with three different forms of CiS stem bark extracts. Analysis of each extract for polyphenolic content revealed the ethanolic solution to have a higher concentration of polyphenols, relative to those of the other aqueous and hydro-ethanolic samples. However, it was interesting to later learn that after the 10-day study period, the aqueous extracts (at doses of 400 mg/kg.bw) significantly decreased the BG compared to the others. In the end, the group was able to confirm the ability of the extract to interfere with both starch and sucrose digestion and recommended that the source be further explored for its potential in the treatment of DM and other metabolic disorders (Boris et al. 2017).

3.24 *CORIANDRUM SATIVUM* LINN. *(CoS)*

Common names: Cilantro, Coriander, Coriandre, Dhania (Figure 3.44)

C. sativum is a herbaceous plant of the Apiaceae or Umbelliferae family and can grow to a height of 0.2–0.6 m. CoS is native to eastern Mediterranean countries, but

FIGURE 3.44 Illustration of *Coriandrum sativum*.

is now cultivated across Europe and America (Pamplona-Roger 2005). The fruit, fresh leaves, and cooked root can be used for various culinary purposes (Waheed et al. 2006); however, its most popular use is as a spice in various food items (Verma et al. 2018). Findings have suggested that only its ripe fruit should be used and never the green parts. Furthermore, the essence of CoS should be consumed in moderation, as high doses could lead to seizures (Pamplona-Roger 2005).

CoS seeds have been shown to reduce the rate of development and the extent of hyperglycaemia in STZ-induced diabetic mice (Bailey and Day 1989). Two constituents of its seed extract that have shown antidiabetic activity have been identified as *p*-cymene (Figure 3.45a) and linalool (Figure 3.45b) (Bailey and Day 1989, Rahimi 2015, Verma et al. 2018).

A clinical investigation by Waheed et al. compared the effects of powdered, aqueous, and alcoholic extracts of CoS on 20 diabetic patients, half of whom had a confirmed history of unsuccessful therapy using prescribed antidiabetic medication. The study group reported that the administration of any of the three forms had resulted in significant decreases in the mean concentrations of PG in the patients and also highlighted the ameliorative effects of the extracts in patients who were previously

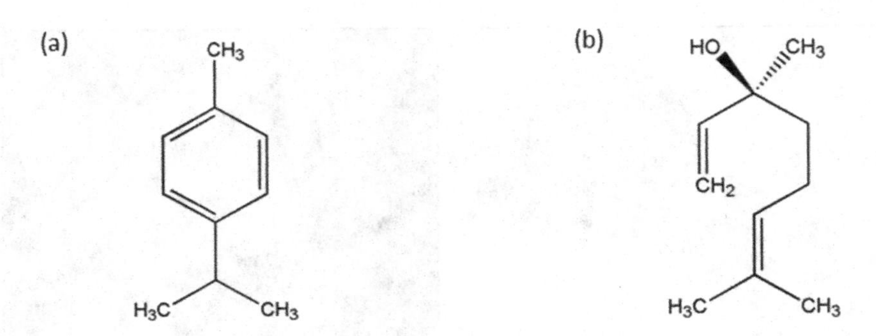

FIGURE 3.45 Structural formulae of isolated compounds from *Coriandrum sativum*: (a) *p*-cymene and (b) linalool.

using allopathic drug therapy, but were unable to control their elevated sugar levels with the prescribed medication (Waheed et al. 2006).

In another study, Eidi et al. experimented with the CoS ethanolic seed extract in a population of STZ-induced diabetic rats. The team reported that a 200 mg/kg.bw dose was able to significantly increase the activity of the pancreatic β-cells and lower the serum glucose levels in the treated diabetic rats, in comparison to their untreated counterparts (Verma et al. 2018, Rahimi 2015, Eidi et al. 2009).

As an alternative mechanism, Das et al. showed that an aqueous extract of the CoS seeds enhanced glucose transport, glucose oxidation, and glycogenesis to an extent that was comparable to 10^{-8} M insulin (Gray and Flatt 1999). However, it was noted that the action of CoS greatly differed from that of metformin (which is still deemed a more potent antidiabetic agent in comparison to the aqueous CoS seed extract), due to its effects on glucose transport via insulin-mediated peripheral glucose uptake (Das et al. 2019).

An *in vivo* study by Aligita et al. showed that the administration of 200, 400, and 800 mg/kg.bw leaf extract doses to different groups of alloxan-induced diabetic mice showed appreciable hyperglycaemic attenuation in comparison to glibenclamide (0.65 mg/kg.bw) over a 14-day period. Interestingly, the group receiving the 200 mg dose showed closer correlation in efficacy to the prescription drug over the other higher doses. The group also performed an *in vitro* study to assess the inhibiting activity of the CoS extract on the AG enzyme (see Chapter 2.5). Based on the experiment's IC_{50} value, the CoS extract showed stronger inhibition of the enzyme (at 32.376 ppm) in relation to the standard drug, acarbose (at 82.272 ppm) (Aligita et al. 2018).

3.25 *CRESCENTIA CUJETE* LINN. *(CrC)*

Common names: Calabacero, Calabash, Kalbas, Miracle fruit, Rum tree, Totumo (Figure 3.46)

Cr. cujete (commonly known as calabash) is a species of fruiting plant that is ubiquitous to Africa, Central America, South America, extreme southern Florida,

FIGURE 3.46 Illustration of *Crescentia cujete.*

and the West Indies. It is dicotyledonous, with simple leaves that are either alternate or clustered on short shoots. In the Caribbean, the calabash fruit is mainly grown for decorative and crockery purposes. However, the revelation of its use in Philippine folk medicine as a treatment for many diseases, including diabetes, warranted its inclusion in this book. The phytochemical composition of CrC has been documented to include alkaloids, iridoids, sterols, triterpenes, peptides (insulin), proteins (bixine), cyanhydric acid (hydrogen cyanide), allicine (Figure 3.8a), pectin (Figure 3.47a), tartaric acid (Figure 3.47b), citric acid (Figure 3.47c), nerolidol (Figure 3.47d), and malic acid (Figure 3.47e), all of which are believed to possess antidiabetic properties. In particular, cyanhydric acid was found to stimulate insulin release, while the alkaloids of CrC were noted to be involved in glycogenesis (Koffi, Édouard and Kouassi 2009). In addition, contained iridoids were noted to reverse high glucose levels and obesity-induced β-cell dysfunction in the pancreas (Zhang et al. 2006), and also stimulate insulin secretion (Sundaram, Shanthi and Sachdanandam 2013). Pectin was shown to increase the activity of glycogen synthase, increase hepatic glycogen, and decrease BG levels (Gomathy, Vijayalekshmi and Kurup 1990, Buchanan, Gruissem and Jones 2000). Furthermore, the presence of citrate (Figure 3.47f) (from citric acid) is found to inhibit phosphofructokinase, which helps regulate glycolysis and, eventually, the citric acid cycle (Murray et al. 2009).

In a study carried out by Uhon et al., the antidiabetic activity of crude ethanolic extracts (CCE) of air-dried CrC leaves was studied. It was reported that the extracts were able to normalize the BG level in a population of 27 alloxan-induced diabetic mice (*Mus musculus*, MM). Furthermore, using ethyl acetate extracts, it was able to mitigate hyperglycaemia in pre-orally treated MM mice. Noteworthy, the team underscored an observed likeness in the effect of the extracts (at concentrations of 10,000 and 5,000 ppm) to that of metformin (Uhon and Billacura 2018).

Similarly, Billacura et al. also gauged the hypoglycaemic activity and protective potential of the CrC fruit against those of metformin. *In vitro* (α-amylase inhibition activity) and *in vivo* screening methods (utilizing MM mice) were performed

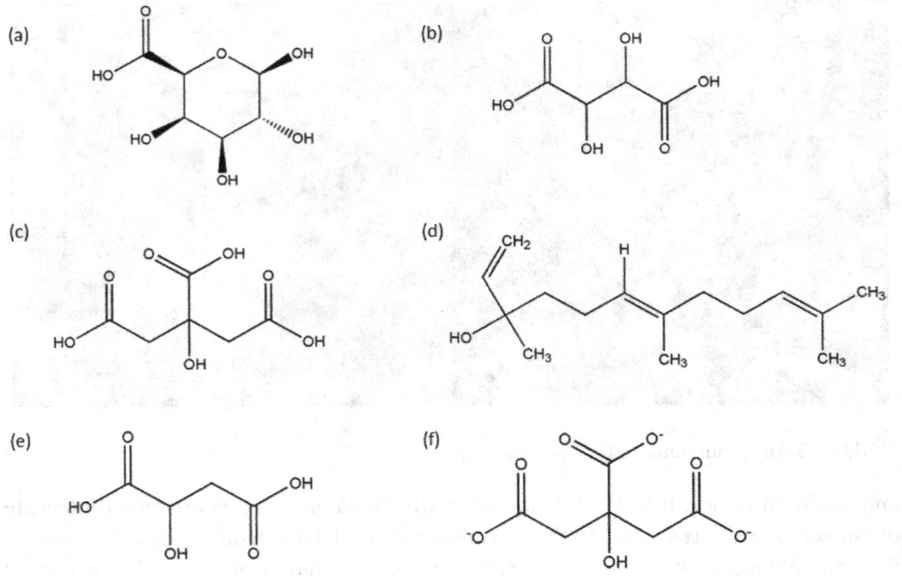

FIGURE 3.47 Structural formulae of isolated compounds from *Crescentia cujete*: (a) pectin, (b) tartaric acid, (c) citric acid, (d) nerolidol, (e) malic acid, and (f) citrate.

using fresh, decocted, hexane, aqueous, and crude ethanolic extracts. Alpha-amylase inhibitory assays revealed that 10,000 ppm hexane extracts were most efficacious in inhibiting the enzyme activity, followed by similar doses of the aqueous and crude ethanolic extracts, respectively. The *in vivo* experiments that followed suggested that the fresh extract and the three solvent extracts (10,000 ppm hexane, 10,000 ppm aqueous, and 10,000 ppm crude ethanolic extract) were also equally capable of lowering elevated BG levels in the diabetic animals. Finally, the group demonstrated that normoglycaemic levels could be maintained in healthy subjects by dietary supplementation with any of the extract forms: fresh (1 : 1 [m/v ratio of the sample : water]), decocted, solvent (hexane [10,000ppm]), aqueous (10,000 ppm), or CCE (5000 or 1000 ppm) (Billacura and Alansado 2017).

3.26 *CUCUMIS SATIVUS* LINN. *(CS)*

Common names: Cucumber (Figure 3.48)

C. sativus belongs to the family Cucurbitaceae and is cultivated on vines that root in the ground, and grow on supporting frames. The leaves are usually large, with the fruit varying in size, shape, and texture. It has been reported that the plant has anti-diabetic properties and scientists have performed several confirmatory studies using animal models, such as rabbits and rats (I. A. Ross 2003).

Sharmin et al. assessed the activity of the aqueous fruit extract on the FBG levels of diabetic rats using a single intraperitoneal injection of the cucumber extract

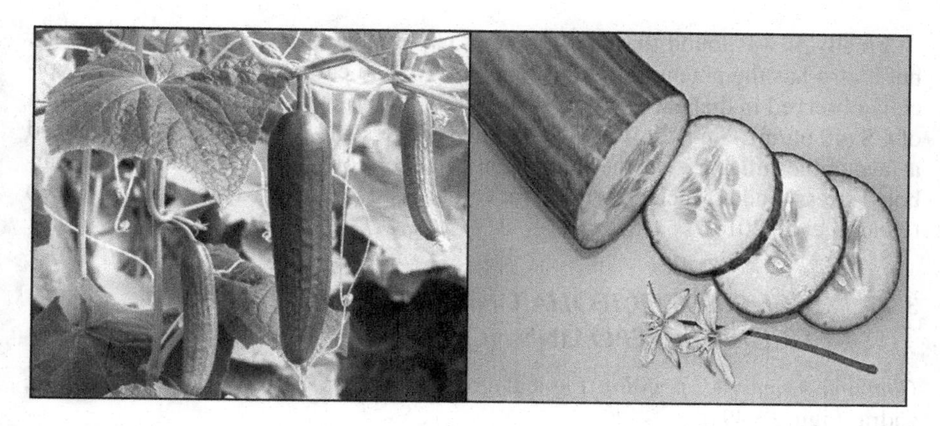

FIGURE 3.48 Illustration of *Cucumis sativus.*

(200 mg/kg.bw). The research group reported good correlation in the extract's activity, by demonstrating a reduction in the rats' FBG levels to 81.02%, 58.65%, and 32.61% at 4, 8, and 12 hours, respectively, in relation to the metformin standard, which showed a reduction of 69.20%, 44.95%, and 25.44% within the same time period (Sharmin et al. 2013).

Ebirien et al. experimented with CS fruit juice on 50 healthy human subjects (aged 18–29 years). The individuals were separated into five groups: Group 1: receiving 200 g CS juice only; Group 2: receiving 200 g CS juice and 200 g rice meal; Group 3: receiving 400 g CS juice only; Group 4: receiving 400 g CS juice and 400 g rice meal; and Group 5: receiving 400 g rice meal only. They discovered that when compared to Groups 1 and 2, Groups 3, 4, and 5 showed a significant difference in PG values between the baseline (pretreatment) individuals and the 2-hour postprandial individuals. From their results, they concluded that the cucumber supplement was able to effect hypoglycaemic activity on the PG levels, since an intake of an increased dose or at concentrations of 400 g caused a significant change in the BG concentration (Bartimaeus, Echeonwu and Ken-Ezihuo 2016).

In another study, Karthiyayini et al. looked at the ethanolic extracts of CS fruit to assess the effectiveness of hyperglycaemic attenuation and hypolipidaemic activity in both healthy glucose-loaded and STZ-induced diabetic rats. To underscore the validity of their study, they included the commercial drug glibenclamide for reference purposes. The group reported that when extracts of 200 and 400 mg/kg.bw of CS were administered to the glucose-loaded healthy rats for 18 hours, a significant reduction in PG levels was observed after 30 minutes. The effect was observed to continue until normal levels were recorded at 90 minutes from the point of administration. They then turned to their diabetic subjects, where they reported that hyperglycaemia was significantly lowered through the administration of 200 and 400 mg/kg.bw of both the CS extract and glibenclamide. Noteworthy was their observation that at 400 mg/kg.bw, the CS extract showed similar activity to that of the commercial drug (Karthiyayini et al. 2009).

Finally, it was found that not only does CS exhibit antihyperglycaemic activity, but it also has the potential to mitigate oxidative and carbonyl stress (which is typically observed in diabetes). In this final study, Heidari et al. used an aqueous extract of CS (40 µg/mL) on freshly isolated rat hepatocytes and proved that the extract had a favourable influence on the cytotoxicity markers of both the oxidative and carbonyl stress models, including those for cell lysis and reactive oxygen species (ROS) (Heidari et al. 2016).

3.27 *CUCURBITA FICIFOLIA* LINN. *(CF) AND CUCURBITA PEPO* LINN. *(CP)*

Common names: *C. ficifolia*: Fig-leaf gourd, Malabar gourd, Black seed squash, Cidra (Figure 3.49a);
C. pepo: Pumpkin, Field pumpkin, Ozark melon, Texas gourd (Figure 3.4b)

The wide variety of pumpkins and squash belong to the Cucurbitaceae family. It is an edible plant that is prepared and consumed in various dishes around the world. Pumpkins are mainly cultivated in temperate, tropical areas and are found growing on vines that trail over the ground or on supporting fences. Its fruits, leaves, seeds, and oil contain active constituents such as alkaloids and flavonoids, which

FIGURE 3.49 Illustrations of *Cucurbita* species: (a) *Cucurbita ficifolia* and (b) *Cucurbita pepo*.

have proven medicinal properties and can be used for antidiabetic, antioxidant, and anti-inflammatory treatments (I. A. Ross 2003).

Xia and Wang demonstrated the similarity in the hypoglycaemic activity of CF with that of the commercial drug tolbutamide, using both healthy animals (with temporary hyperglycaemia) and mildly diabetic animals. Extracts from the pumpkin fruits (300 mg/kg.bw) were administered daily (over a period of 30 days) to STZ-induced diabetic rats, during which the BG levels of the subjects were routinely monitored. The results indicated that feeding with the fruit extract caused reductions in STZ-induced hyperglycaemia, while simultaneously increasing the plasma insulin levels and markedly reducing the STZ-induced lipid peroxidation in the pancreas. There was also a significant increase in the number of pancreatic β-cells in the CF-treated animals when compared to the untreated diabetics. However, it should be noted that the degree of β-cell restoration of the CF-treated rats was not as high as that of the healthy rats, thus hinting the degree of potency of the CF fruit extract (Xia and Wang 2007).

In another study, Kwon et al. investigated the antidiabetic and antihypertensive potentials of the phenolic phytochemicals contained in CP. Interestingly, the study revealed that there was a variation in the bio-active phenol concentrations in relation to the species of pumpkin tested. Among the varieties of pumpkin extracts evaluated, round/orange and spotted orange/green had the highest content of total phenolics and exhibited moderate antioxidant activity, coupled with moderate to high glucosidase and angiotensin I-converting enzyme (ACE) inhibitory effects (Kwon et al. 2007).

Finally, Makni et al. looked at the ameliorative effects of the compounds contained in a powdered seed mixture of pumpkin (CP) and flax seeds (*Linum usitatissimum* Linn.) on a test group of alloxan-induced diabetic rats. Here, they found that upon administration of the seed mixture, there was an observed decrease in the BG concentration in the treated rats. Furthermore, continued administration of the feed mixture until day-1 resulted in a 63% increase in plasma insulin concentration, as well as an increase in antioxidant effects.

3.28 *CURCUMA LONGA* LINN. *(CL)*

Common names: Curcuma, Haldi, Hardi, West Indian saffron, Safflower, Turmeric, Yellow Ginger (Figure 3.50)

The spice turmeric that is derived from the root of the plant *C. longa* has been described as a treatment for diabetes in Ayurvedic medicine for thousands of years (Aggarwal et al. 2007). In a very comprehensive review carried out by Zhang et al., the many studies and benefits of CL were discussed and summarized (Zhang et al. 2013). Curcumin (Figure 3.51) has been identified as the predominant active component of turmeric and has caught the attention of the scientific community as a potential therapeutic agent in both the direct treatment of diabetes and the treatment of its associated complications (Pérez-Torres et al. 2013). This popularity is well-deserved because of its effectiveness in reducing hyperglycaemia and hyperlipidaemia in rodent models and, of equal importance, its relatively low cost and low toxicity

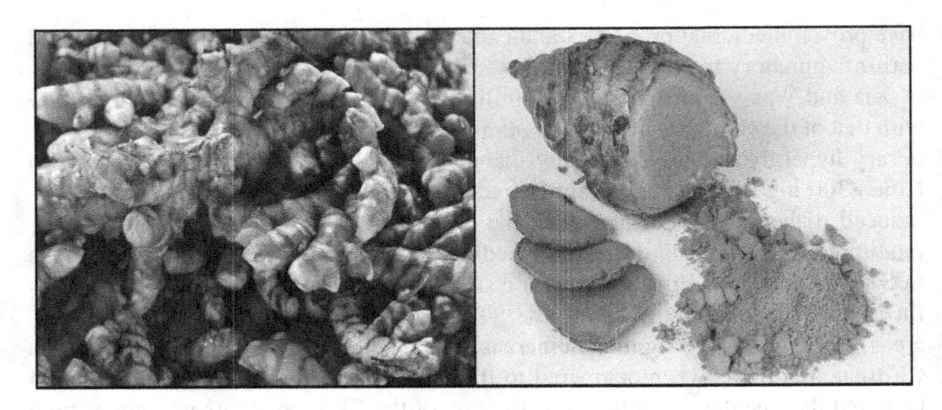

FIGURE 3.50 Illustration of *Curcuma longa*.

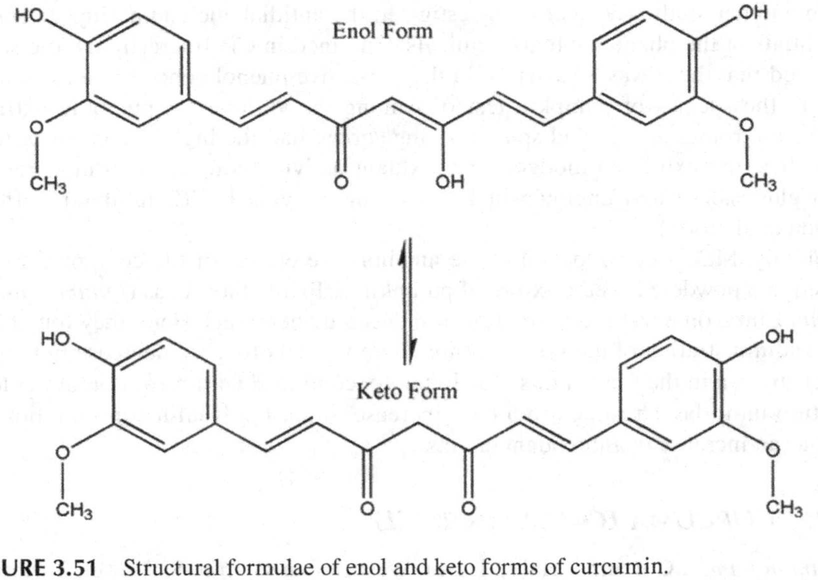

Enol Form

HO OH O OH O CH_3 CH_3

Keto Form

HO OH O O O O CH_3 CH_3

FIGURE 3.51 Structural formulae of enol and keto forms of curcumin.

(Goel, Kunnumakkara and Aggarwal 2008, Shehzad et al. 2011, Chuengsamarn et al. 2012, Sahebkar 2013, Evangelopoulos et al. 2014).

In diabetes-induced rat models (Pari and Murugan 2007), the oral administration of various dosages of curcumin (ranging from 80 to 300 mg/kg.bw, during a 2 to 8-week period) (Pari and Murugan 2007, Arun and Nalini 2002, Murugan and Pari 2007, Soetikno, Watanabe et al. 2011, Xavier et al. 2012, Na et al. 2011, Patumraj, Wongeakin et al. 2006, Soetikno, Sari et al. 2011, Jain et al. 2009) was able to prevent body weight loss, reduce the levels of glucose, haemoglobin (Hb), and HbA1c in blood (Arun and Nalini 2002), and improve insulin sensitivity (Murugan and Pari 2007). Additionally, the oral administration of turmeric aqueous extract (300 mg/kg.

bw) (Hussain 2002) or curcumin (30 mg/kg.bw) for 56 days (Mahesh, Balasubashini and Menon 2004) resulted in a significant reduction in the BG in STZ-induced diabetic rat models. In high-fat diet (HFD)-induced insulin resistance and T2DM models in rats, the oral administration of curcumin (80 mg/kg.bw) for 15 and 60 days, respectively, showed an antihyperglycaemic effect and improved insulin sensitivity (El-Moselhy et al. 2011). Dietary curcumin (0.5% in diet) was also effective in ameliorating the increased levels of FBG, urine sugar, and urine volume in STZ-induced diabetic rats (Chougala et al. 2012). Unfortunately, despite all these and many more studies that support the effectiveness of CF therapy, the data is still criticized for not being "high-quality clinical evidence" for its use in treating any disease (Nelson et al. 2017)

3.29 *CYMBOPOGON CITRATUS* LINN. *(CC)*

Common names: Citron grass, Citronela, Citronella grass, Fever grass, Lemon grass (Figure 3.52)

Cy. citratus (also known as *Andropogon citratus* Linn.) is a perennial plant of the Poaceae or Gramineae family (Shah et al. 2011). It is cultivated in temperate and tropical regions and is native to Asia, Africa, and America (Garba et al. 2020, Machraoui et al. 2018, Oladeji et al. 2019). It is a grass-like plant that can grow up to 2 m in height, with a plant width of approximately 1.2 m (Oguntibeju 2019, Machraoui et al. 2018).

Various reports have shown the extracts of CC to exhibit desirable hypoglycaemic effects (Shah et al. 2011, Garba et al. 2020, Machraoui et al. 2018, Abbas et al. 2018).

FIGURE 3.52 Illustration of *Cymbopogon citratus.*

Analysis of these extracts have revealed an assemblage of different bioactive compounds, which when purified have yielded the most efficacious isolates: citral (existing as two isomers: gerenial and neral) (Figure 3.53a) and β-myrcene (Figure 3.53b) (Mishra et al. 2019, Omari et al. 2007). It is important to note that although the aqueous extract of the leaves is more commonly used, similar extracts from the roots and flowers were observed to display greater antidiabetic activity than its co-candidate source (Abbas et al. 2018).

Oguntibeju et al., Adeneye et al., and Sharma et al. independently explored the effects of CC leaf extracts on diabetic rats, and unanimously confirmed their efficacy in attenuating the elevated BG levels in the subjects (Oguntibeju 2019, Adeneye and Agbaje 2007, Sharma and Gupta 2017). Another study by Ademuyiwa et al., was undertaken to investigate the effect of different forms of extracts on the blood sugar level, lipid profiles, and hormonal profiles of normal rats. Reports showed that the oral administration of the ethanolic and aqueous forms (200 mg/kg.bw for 30 days) caused a significant reduction in BG levels. Furthermore, the thyroid-stimulating hormone (TSH), triiodothyronine (T3), and thyroxine (T4) hormonal levels were found to be desirably higher in all the administered groups, while the lipid profile levels were observed to be lowered in the treated rats (Ademuyiwa, Olamide and Oluwatosin 2015).

Alternatively, Bharti et al. experimented with the essential oil of CC and found that it significantly lowered the insulin resistance of diabetic rats (evaluated by the homeostatic model assessment for insulin resistance [HOMA-IR] index). The group observed a dose-response effect to the therapy, as the drop in insulin resistance was

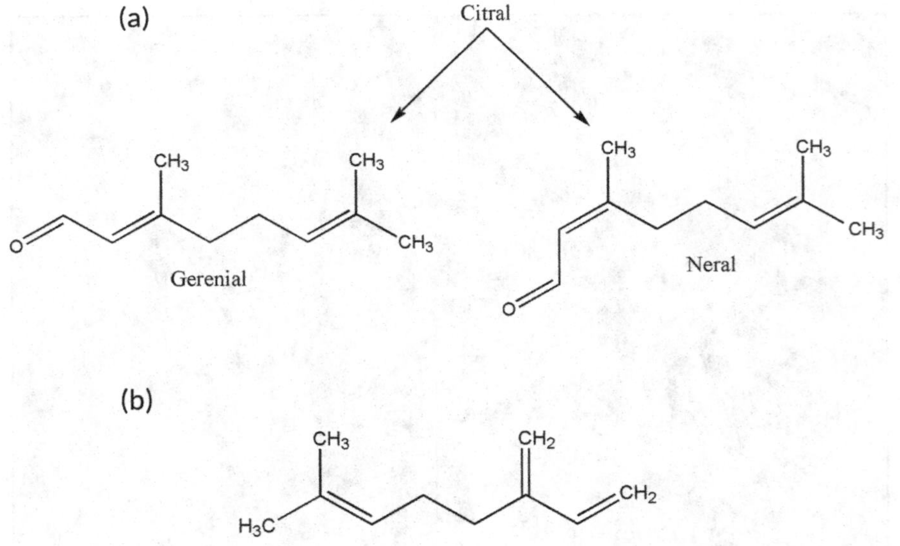

(a)

Citral

CH_3 CH_3 O CH_3

Gerenial

CH_3 CH_3 CH_3 O

Neral

(b)

CH_3 CH_2 CH_2 H_3C

FIGURE 3.53 Structural formulae of isolated compounds from *Cymbopogon citratus*: (a) citral and (b) β-myrcene.

more evident in the diabetic rats, which were treated with doses of 800 mg/kg.bw compared to those treated with 400 mg/kg.bw (Oladeji et al. 2019, Bharti et al. 2013). Similarly, the Buaduo-led group reported satisfactory α-glucosidase and α-amylase inhibitory effects of CC extracts, thus confirming its use as a potent regulator in glucose metabolism (Boaduo et al. 2014).

While promising in its efficacy, toxicity studies on the extracts of CC have produced quite conflicting results. Formigoni et al. found that the administration of high-level doses of an infusion of CC leaves to rats was deemed non-toxic (Formigoni et al. 1986). However, another study revealed that at higher doses of 1500 mg/kg.bw, CC oils caused significant functional damage to the stomach and liver of Wistar rats, and at a dose of 2000 mg/kg.bw, tea from CC dried leaves also caused similar damage (Fandohan et al. 2008). This again underscores the need for more robust toxicity studies to be performed on this and other biologically active plant extracts to gauge their safety in use.

3.30 *EUPHORBIA HIRTA* LINN. *(EH)*

Common names: Asthma plant, Asthma weed, Soro Soro, Garden spurge, Snakeweed (Figure 3.54)

E. hirta (also known as *E. pilulifera* Linn.) is a small, erect plant of the Euphorbiaceae family that reaches an approximate height of 0.5 m (Al-Snafi 2017). It is native to India and Australia but is now widely found in both tropical and subtropical countries (N'Guessan Bra Fofie et al. 2013).

The presence of numerous chief chemical constituents in EH, including quercitrin (Figure 3.55a), polyphenolic flavonoids, and palmitic acid (Figure 3.55b), has supported its potential use in T2DM therapy (Sharma et al. 2018, Le et al. 2018).

In this first study, Kumar et al. reported on the effects of ethanolic and petroleum-ether extracts of the stems and concluded that the respective extracts had impactful antihyperglycaemic activity (Kumar, Rashmi and Kumar 2010, Kumar, Malhotra

FIGURE 3.54 Illustration of *Euphorbia hirta.*

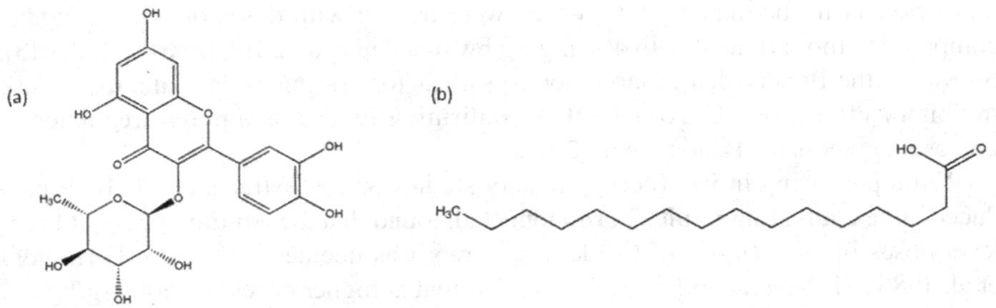

(a) (b)

FIGURE 3.55 Structural formulae of isolated compounds from *Euphorbia hirta*: (a) quercitrin and (b) palmitic acid.

and Kumar 2010). Supplementary explorations showed that not only the stem but also the ethanolic extracts of the leaf, flower, and root had therapeutic effects on elevated BG levels (Al-Snafi 2017, Kumar, Rashmi and Kumar 2010).

The investigating teams also determined that the antidiabetic mechanism of both ethanolic and ethyl acetate extracts followed an α-glucosidase inhibitory pathway (Al-Snafi 2017, Sharma et al. 2018, Widharna et al. 2010). Alternatively, a study carried out by Subramanian et al. suggested that the antidiabetic activity of the EH plant extract was due to an insulin stimulatory effect from remnant β-cells (Subramanian, Bhuvaneshwari and Prasath 2011). In yet another direction, *in vitro* investigations performed by Shilpa et al. (using the methanolic extract of the whole plant) proposed another mechanism of action, in which the active compounds served as inhibitors to α-amylase activity (Shilpa et al. 2020). Adding to this, independent investigations using similar extract forms showed that doses of 200 and 400 mg/kg .bw influenced striking decreases in the BG levels, and also gave some protection and recovery from dyslipidaemia and oxidative stress in STZ-induced diabetic rats (Devi and Kumar 2017).

Regarding its associated toxicity, it was indeed encouraging to note that complimentary to its reported desirable activities, (Al-Snafi 2017, Sharma et al. 2018, Subramanian, Bhuvaneshwari and Prasath 2011, Uppal, Nigam and Kumar 2012, Maurya et al. 2012), no cytotoxic events were observed (in doses up to 500 mg/kg.bw) (Uppal, Nigam and Kumar 2012) and no mortalities (in doses up to 2000 mg/kg.bw).

3.31 *HIBISCUS ROSA-SINENSIS* LINN. *(HRS)*

Common names: Hibiscus, Rose mallow, Shoe black plant, Shoe flower plant, Tulipan (Figure 3.56)

H. rosa-sinesis is a perennial, ornamental, glabrous shrub of the Malvaceae family (Upadhyay and Upadhyay 2011, Bhaskar and Gopalakrishnan 2012) which is native to Southeast Asia, but is now mainly grown in tropical regions (I. A. Ross 2003). Due to their observed hypoglycaemic effect, extracts of both the flowers (Upadhyay and Upadhyay 2011, Pillai and Mini 2016, Venkatesh, Thilagavathi and Shyam sundar 2008, Ghosh and Dutta 2017) and the leaves (I. A. Ross 2003, Ghosh

FIGURE 3.56 Illustration of *Hibiscus rosa-sinesis.*

and Dutta 2017, Sachdewa and Khemani 2003, Mamun et al. 2013) of the plant have been commonly considered for diabetes treatment (Moqbel et al. 2011). Additionally, a few recent studies have suggested that the root extract can also be used as a viable therapeutant (Kumar et al. 2013).

Several studies have demonstrated the glucose-lowering effect of the ethanolic flower extracts (at common doses of 250 and 500 mg/kg.bw) in diabetic rats (Bhaskar and Gopalakrishnan 2012, Venkatesh, Thilagavathi and Shyam sundar 2008, Ghosh and Dutta 2017). Adding to this, Ghosh and Dutta observed a boost in the insulin levels of their test animals, when extracts of a similar nature were used in their study (Ghosh and Dutta 2017). Of particular importance, Sachdewa and Khemani found that the flower extracts influenced a significant and comparable BG attenuating effect to that of the commercial drug, glibenclamide (Bhaskar and Gopalakrishnan 2012, Sachdewa and Khemani 2003).

Alternatively, there were also significant improvements in the ability to utilize the external glucose load, when the alcoholic leaf extract was orally administered to rats at a dose of 250 mg/kg.bw daily for 7 days (I. A. Ross 2003). Moqbel et al. reported that the oral administration of the fractionated ethanolic leaf extract (at a dose of 200 mg/kg.bw) to non-obese diabetic mice, exhibited the greatest ameliorative insulinotropic effect in relation to the other retrieved fractions (Moqbel et al. 2011). Sachdewa et al. also investigated the effect of the aqueous leaf extract and found good correlation with the control standard, tolbutamide (Sachdewa, Nigam and Khemani 2001).

Regarding its toxicity or its influence on the occurrence of toxic events, we found that the intraperitoneal administration of a 70% ethanolic leaf extract to mice exhibited an LD_{50} of 1.533 g/kg.bw. Similarly, additional rodent studies showed that the intraperitoneal administration of an ethanolic extract solution of the aerial parts of the plant showed an LD_{50} of 1 g/kg.bw (I. A. Ross 2003). From these studies, we have noted that the effective doses that were used in the aforementioned studies were well below the reported lethal doses and as such, the use of the HRS extract can be deemed a relatively safe form of therapy.

3.32 *JUSTICIA SECUNDA* VAHL *(JS)*

Common names: Saint John's bush (Figure 3.57)

J. secunda is a vibrant red flowering plant belonging to the Acanthaceae family that grows in the tropical and subtropical zones of Africa, Asia, Madagascar, and the Caribbean (Mea et al. 2017). In Caribbean traditional medicine, herbal tea is made from the plant that is consumed for menstrual pain (Useful Tropical Plants n.d.). The leaves and stems are used in the treatment of anaemia, coughs, colds, fevers, and amenorrhoea, while (Useful Tropical Plants n.d., DeFilipps, Maina and Crepin 2004) in African pharmacopoeia, JS is used for its antidiabetic properties (Mea et al. 2017).

Mea et al. investigated the possible hypoglycaemic effects of JS on diabetic rats. In their study, 2.5 and 3 g/kg.bw of an aqueous extract of JS leaves were administered to the test animals. After the treatment, they found that both doses significantly decreased BG levels over time; however, a lower reduction was observed at the 3 g/kg.bw dose (Mea et al. 2017).

In another study, Theiler et al. proposed that the compounds contained in the JS leaf extract possessed α-glucosidase-inhibiting capabilities, which as a consequence, could influence a decrease in patient BG levels. From their investigations, not only was the group able to confirm the enzyme inhibiting efficacy of the extracts, but it was also successful in isolating and characterizing two novel compounds that seemed to be responsible for the activity: Compound 1 (Figure 3.58a) and Compound 2 (Figure 3.58b) (Theiler et al. 2016).

FIGURE 3.57 Illustration of *Justicia secunda.*

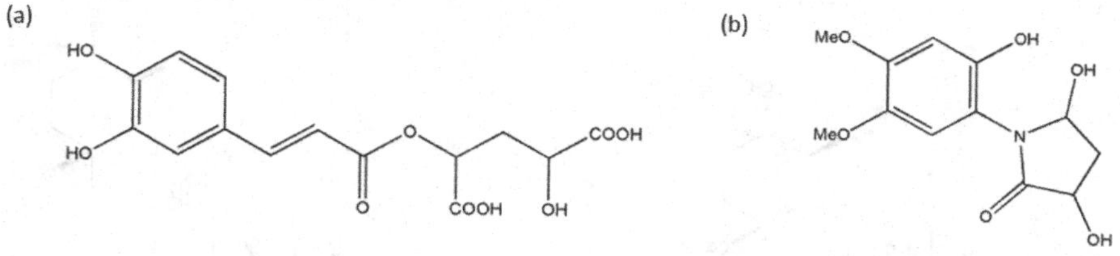

FIGURE 3.58 Structural formulae of isolated compounds from *Justicia secunda*: (a) Compound 1 and (b) Compound 2.

3.33 *LANTANA CAMARA* LINN. *(LC)*

Common names: Common Lantana, Lantana, Red Sage, Shrub Verbena, Yellow Sage, Caraquite, Cariaquito, West Indian lantana (Figure 3.59).

L. camara is an erect, hairy, aromatic shrub that grows to an approximate height of 3 m. The plant belongs to the Verbenaceae family, and can be found in tropical and subtropical regions, such as parts of the Americas, Asia, and Africa. LC contains significant amounts of essential oils, with higher concentrations found in the flowers and leaves rather than other parts, such as the roots and stem (Nawaz et al. 2016, Kalita et al. 2012).

Qualitative screening and subsequent antidiabetic evaluations have been done on the phytochemicals contained in the foliar extract of LC. It was reported that the presence of common phenols and flavonoids was responsible for its α-amylase inhibiting activity, in a dose-dependent manner (Bhowmick, Bartariya and Kumar 2018). Furthermore, two triterpenoid compounds (urs-12-en-3β-ol-28-oic acid [a stearoyl glucoside of ursolic acid] [Figure 3.60a] and urs-12-en-3β-ol-28-oic acid 3β-D-glucopyranosyl-4′-octadecanoate [Figure 3.60b]) have been isolated and identified to be predominantly responsible for the plant's antidiabetic activity (Nawaz et al. 2016, Kazmi et al. 2012). Supporting this, Jawonisi and Adoga provided independent evidence of the antidiabetic

FIGURE 3.59 Illustration of *Lantana camara*.

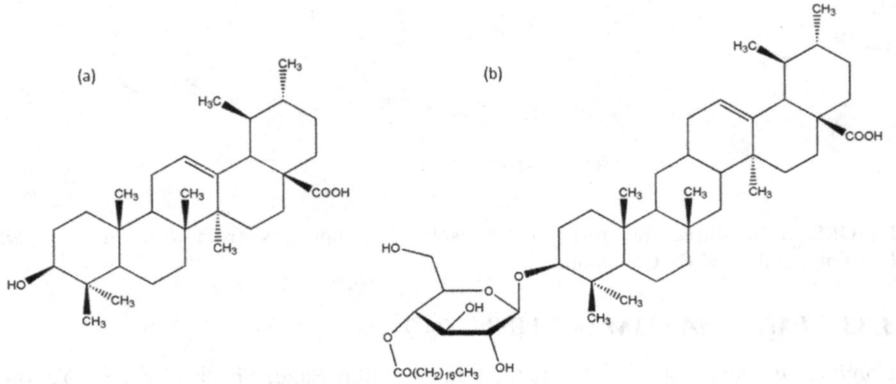

FIGURE 3.60 Structural formulae of isolated compounds from *Lantana camara*: (a) Urs-12-en-3β-ol-28-oic acid and (b) urs-12-en-3β-ol-28-oic acid 3β-D-glucopyranosyl-4′-octadecanoate.

activity of the LC leaf extract constituents and underscored the therapeutic potential of this class of natural products (Jawonisi and Adoga 2015).

Additionally, it was observed that the methanolic extracts of LC leaves and fruits also showed antihyperglycaemic activity (Kalita et al. 2012, Pramanik and Giri 2018, Sangeetha, Mahendran and Ushadevi 2015). Complementary to this, Venkatachalam et al. administered oral doses of 100 and 200 mg/kg.bw (methanolic fruit extract) to STZ-induced diabetic rats, and confirmed its antihyperglycaemic effects (Venkatachalam et al. 2011). Another investigating team confirmed the antidiabetic activity of the plant, using 200 and 400 mg/kg.bw doses (aqueous leaf extract) on alloxan-induced hyperglycaemic rats (Dash, Suresh and Ganapaty 2001). We found comparable results in other studies, where the methanolic extracts (at the same doses) were administered to both STZ-induced (Loganathan and Bhadauria 2019), and alloxan-induced diabetic rats (Ganesh et al. 2010). Khatoon et al. supportively showed that there was similar normalization of BG levels, using the ethanolic leaf extract, at doses of 250 and 500 mg/kg.bw (Khatoon et al. 2013).

Before concluding, it is very important to note that LC is considered one of the most toxic plants (Kalita et al. 2012, Dash, Suresh and Ganapaty 2001) however, the occurrence of toxic events has only been reported following a high consumption of the plant material (Kalita et al. 2012). Encouragingly, Venkatachalam et al. stated that their acute toxicity studies revealed that the methanolic extract of the LC fruits was non-toxic (Venkatachalam et al. 2011), and Khatoon et al. claimed that in their study, the diabetic rats treated with the ethanolic leaf extract (at a dose of 2000 mg/kg.bw) did not exhibit any toxicity-associated side effects (Khatoon et al. 2013).

3.34 *LEONOTIS NEPETIFOLIA* LINN. *(LN)*

Common names: Klip dagga, Christmas candlestick, Lion's ear, Chandilay bush (Figure 3.61)

FIGURE 3.61 Illustration of *Leonotis nepetifolia*.

L. nepetifolia is a plant belonging to the Lamiaceae family and is found in tropical and subtropical regions such as Africa, Asia, and the Caribbean. Although they are not traditionally eaten as food, all parts of the plant are known to possess medicinal properties, due to their wide spectrum of chemical constituents (Gungurthy et al. 2013). In Trinidad, it is used as a traditional medicine (infusion) against fever, coughs, and malaria (Gungurthy et al. 2013, Mendes 1986, TROPILAB® INC. n.d.); however, little research has been done to evaluate its antidiabetic activity.

We were able to find a study that was conducted by Gungurthy et al., which showed promise for the plant to be used in this aspect. Preliminary phytochemical analysis indicated the presence of glycosides, flavonoids (quercetin: Figure 3.25) and alkaloids, carbohydrates, steroids and terpenoids, proteins, fixed oils and volatile oils, and phenolics and tannin, all of which have been previously discussed as being responsible for the potent hypoglycaemic effects in other botanical sources (I. A. Ross 2003).

Results from their oral glucose tolerance test (OGTT), which was done to assess the effect of varying concentrations of the extract and also to gauge their respective efficacies and potencies against the control standard: glibenclamide, showed an appreciable and steady decrease in the BG levels, with the most comparative to the control drug being the ethanolic extract, at 500 mg/kg.bw (Gungurthy et al. 2013).

In their *in vivo* studies,. Twenty-four alloxan-induced diabetic rats were equally divided into four groups (Groups B–E). Treatment of the groups was as follows: normal saline solution was administered to Groups A (healthy rats) and B (diabetic control), while LN ethanolic extracts (500 and 250 mg/kg.bw) were orally administered to Groups D and E, respectively. Finally, Group C animals received the reference drug glibenclamide (10 mg/kg.bw) and used as the control standard. Blood samples were then collected from the rat tail vein and BG levels were estimated using a One Touch® glucometer on Days 1, 7, 14, and 21. At the end, it was observed that the BG levels of the rats belonging to Groups D and E had dropped from their initial hyperglycaemic level of ≥250 mg/dL. However, the greatest degree of reduction was observed for the subjects of Group D (Gungurthy et al. 2013).

While the results of this study seem very promising, further research is necessary to firstly properly isolate and identify the efficacious compounds, and secondly, to properly evaluate the relative toxicities of their respective dosages.

3.35 *MANGIFERA INDICA* LINN. *(MI)*

Common names: Aam, Manako, Mango (Figure 3.62)

M. indica is one of the oldest cultivated plants in the world, having been grown in India for over 4000 years, and subsequently introduced to the Caribbean during the colonization of the Western world. Belonging to the family: Anacardiaceae, the use of some 30 different species of mangoes goes well beyond its original purpose of food. Different parts of the fruiting tree have been found to possess medicinal properties that can be used for antidiabetic, antioxidant, and anti-inflammatory applications (Verma et al. 2018).

Shah et al. investigated the antidiabetic activity of the constituent compounds found in mango leaves and was able to confirm the hypoglycaemic activity of the isolated compound: mangiferin (Figure 3.63a) (Shah et al. 2010).

Furthermore, studies have found that α-carotene (Figure 3.63b) and β-carotene (Figure 3.63c) are converted to vitamin A (Figure 3.27a), which, in turn, stimulates the insulin secretion of the β-cells in the pancreas (Verma et al. 2018, Perry et al. 2009). These active constituents promote a reduction in the intestinal absorption of glucose in the small intestine, which leads to hypoglycaemic activity in the blood.

FIGURE 3.62 Illustration of *Mangifera indica*.

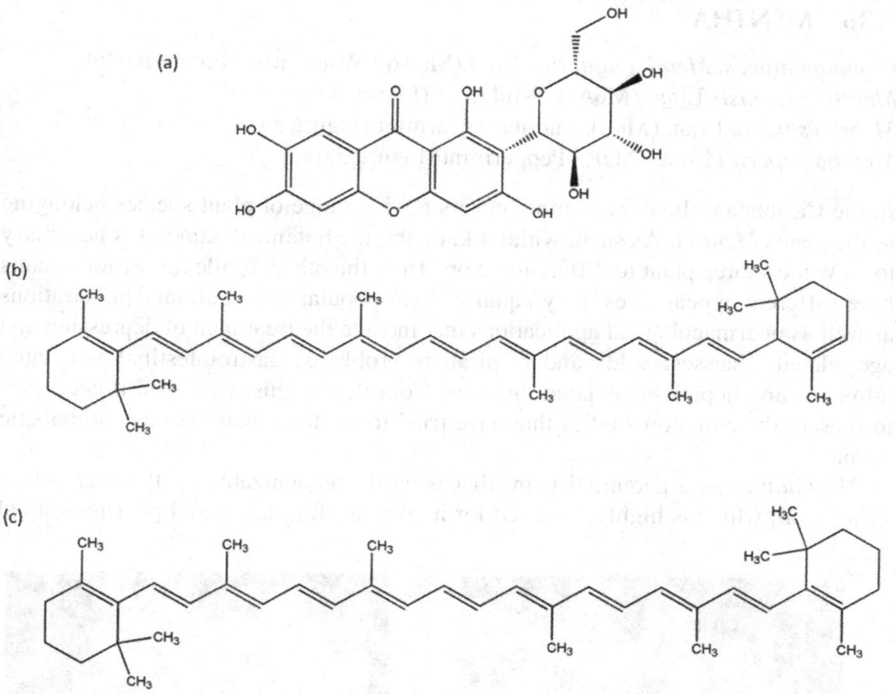

(a)

(b)

(c)

FIGURE 3.63 Structural formulae of isolated compounds from *Mangifera indica*: (a) mangiferin, (b) α-carotene, and (c) β-carotene.

Bhowmik et al. compared the antihyperglycaemic effect of ethanolic extracts of the MI stem, barks, and leaves on T2DM model rats. The group reported that both extracts exhibited significant antihyperglycaemic activity in the subjects, when fed simultaneously with a glucose load. Noteworthy was the observation that the ethanolic extract showed superior efficacy to the aqueous extract (being able to significantly reduce BG levels, when administered 30 minutes prior to the glucose load) (Bhowmik et al. 2009).

In another study, Sánchez et al. reported that an administered dose of 1 mg/kg.bw aqueous extract (60 minutes before glucose loading) was able to attenuate hyperglycaemic levels in rats (Sánchez et al. 2000). Other studies also revealed that the ethanolic extract from the MI bark was also found to display appreciable α-glucosidase inhibitory activity, thus enabling postprandial glucose regulation (Prashanth et al. 2001).

Additional investigations, conducted by Adedosu et al. (Adedosu et al. 2018) and Mohammed et al. (Luka and Mohammed 2012) have furnished further supportive evidence of the antidiabetic activity displayed by MI extracts. Despite all the supportive data however, it should be noted that the respective mechanisms of action of the plant extract are yet to be confirmed.

3.36 MENTHA

Common names: *Mentha aquatica* Linn.(MeAq): Water mint (Figure 3.64a);
Mentha arvenisis Linn. (MeAr): Wild mint (Figure 3.64b);
Mentha spicata Linn. (MeS): Pudina, Spearmint (Figure 3.64c);
Mentha peperita Linn. (MeP): Peppermint (Figure 3.64d)

In the Caribbean, the name "mint" covers a wide range of plant species belonging to the genus *Mentha.* As such, without knowing the botanical name, it is necessary to view the source plant to differentiate one from the other. While the various plants have different appearances, they equally share popularity in culinary preparations as well as pharmacological applications that include the treatment of depression and age-related illnesses, colds and respiratory problems, gastrointestinal-associated ailments, and hepatic-associated diseases. Considering this, we found it necessary to present the common studies that have tried to confirm their uses as antidiabetic agents.

M. aquatica is a perennial plant that is easily recognizable by the scent of its volatile oil, which is highly favoured for its flavour, fragrance, and pharmaceutical

FIGURE 3.64 Illustrations of *Mentha* species: (a) *Mentha aquatica*, (b) *Mentha arvensis*, (c) *Mentha spicata*, and (d) *Mentha piperita.*

applications. Regarding the latter application, the plant was reported to exhibit lipid peroxidative, antimicrobial, antioxidant, and anti-inflammatory activities. Konda et al. proposed to evaluate the antidiabetic and nephron-protective activities of MeAq in STZ-induced diabetic rats. The results showed that the oral administration of its aqueous leaf extract (at a dose of 100 mg/kg.bw/day for 90 days) significantly decreased the level of FBG, HbA1c, TC, TG, plasma urea, creatinine, urine albumin, and LPO. Furthermore, the rats were observed to have increased body weight, insulin, high-density lipoprotein (HDL) cholesterol, plasma albumin, urinary urea, urinary creatinine, and antioxidant enzyme activities. In conclusion, the antidiabetic and nephron-protective potential activities of the extract were attributed to the presence of the plant's phytoconstituents and their synergistic effects (Konda et al. 2020).

Next, we look at the antidiabetic activity of *M. arvensis*. Agawane et al. used the methanolic extract of the MeAr leaves to evaluate its antioxidant and antiglycation potential, through both *in vitro* and *in vivo* studies. Here, 50 mg/kg.bw (methanolic whole plant extract) showed remarkable inhibitory effects on the key enzymes:α-amylase and α-glucosidase(which are linked to T2DM), and influenced a significant reduction in postprandial hyperglycaemia in starch-induced diabetic Wistar rats. From the study, it was suggested that the tested extract had the potential to emerge as a means of treating postprandial hyperglycaemia (Agawane et al. 2017).

M. spicata is a rhizomatous, perennial plant with a height range of 30–100 cm. It has square-shaped, hairy stems and foliage and a wide-spreading fleshy underground rhizome. The leaves are 5–9 cm long and 1.5–3 cm broad, with a serrated margin. Like its related members within the genus, MS owes its broad pharmacological uses to the flavonoids (Figure 3.4), terpenoids, and phenols that have been isolated from its different extracts.

In a study by Bayani et al., the hypoglycaemic activity of MeS was gauged against that of the sulfonylurea, glibenclamide (2 mg/kg.bw). Initially, diabetes was induced in male rats by a single intraperitoneal injection of alloxan monohydrate (150 mg/kg .bw). Therapy then followed, with the oral administration of the aqueous extract (at a dose of 300 mg/kg.bw) over a period of 21 days. The results revealed that the extract displayed similar efficacies to those of the commercially marketed drug and was able to produce a significant reduction in FBG, TC, TG, LDL, and serum malondialdehyde (Bayani, Ahmadi-hamedani and Javan 2017).

M. piperita (peppermint) is known to be a hybrid species, arising from a cross between water mint and spearmint. Its antidiabetic activity was studied by two independent research groups, led by Angel and Barbalho, respectively. In the first study, the oral administration of peppermint juice (over a 21-day period at an extracted concentration of 100 g/L) produced a significant decrease in the BG level of 30 alloxan-induced diabetic rats (Angel, Sai Sailesh and Mukkadan 2013). Such results were mirrored by the Barbalho-led team, who although using similar extract concentrations as the Angel group, performed their study on an F1 population of male inbred Wistar rats. The results showed that the offspring of the treated diabetic females exhibited significantly reduced levels of glucose, cholesterol, LDL, and triglycerides, and also displayed a significant increase in their HDL levels. In their conclusion, it was the group's opinion that the use of MeP juice was a culturally appropriate strategy to aid in the prevention of DM, dyslipidaemia, and its associated complications (Barbalho et al. 2011).

3.37 *MIMOSA PUDICA* LINN. *(MP)*

Common names: Chhui-mui, Humble plant, Lajja, Laajvanti, Teemarie, Ti-Marie, Touch-me-not plant, Sensitive plant, Shameful plant (Figure 3.65)

M. pudica is recognized as a creeping annual or perennial plant, belonging to the Fabaceae family. Descended from parts of Central and South America and southern India, this prickly plant has quite a uniqueness: specifically its bending movement, giving it its curiosity status (Ahmad et al. 2012). It responds to environmental stimuli, hence one of its names "Touch-me-not." Two well-known movements exhibited by MP are a very rapid movement of the leaves when it is stimulated by touch, heat, etc., and a very slow movement of the leaves controlled by a biological clock (Ueda and Yamamura 1999). MP is known to possess anti-asthmatic, antidepressant, tonic, and sedative properties due to the presence of alkaloids, flavonoids, and tannins (Ahmad et al. 2012). In India, the use of MP as a natural treatment for T2DM has also been noted.

Tunna et al. conducted a study on the antidiabetic potential and inhibitory activities of its methanolic extract. The mentioned crude extract underwent sequential fractionation with hexane, ethyl acetate, acetone, and methanol successively, thus furnishing the purer, more soluble components. Using in-depth chemical profiling (gas chromatography and quadrupole time of flight mass spectrometry to assess the phenolic compounds), it was determined that both the methanol and acetone fractions had the highest inhibitory activities against α-amylase and α-glucosidase enzymes (Tunna et al. 2015).

Sutar et al. also compared the antidiabetic activity of the leaf extract of MP in alloxan-induced albino diabetic rats against that of the commercial drug, metformin. The rodents were divided into four different groups: Group 1: control group;

FIGURE 3.65 Illustration of *Mimosa pudica*.

Group 2: diabetic rats (treated with metformin [500 mg/kg.bw/day]); Group 3: diabetic rats (treated with a suspension of petroleum ether extract [600 mg/kg.bw /day]); and Group 4: diabetic rats (treated with a suspension of ethanolic extract [600 mg/kg.bw/day]). The subjects were then administered the extracts and reference drug daily, for a total of 7 days. On the first day, the ethanolic extract showed a significant decrease in serum glucose level to 32.46% as compared to metformin, which was 43.57% at the fifth hour. This trend continued to Day 7, when the ethanolic extract further reduced serum glucose to 50.35% as compared to metformin (62.44%). Unfortunately, the petroleum ether extract did not show any significant reduction in the serum glucose level. From the results, the group was able to confirm that the active constituents responsible for the plant's antidiabetic properties were selectively present in the ethanolic extract and could be isolated in such a manner (Sutar, Sutar and Behera 2009).

Sundaresan and Radhiga also studied the effect of MP leaf extract on high fructose diet–induced T2DM rats. The rats were divided into four groups: Group 1: normal untreated rats; Group 2: normal rats (administered MP at a dose of 100 mg/kg.bw orally twice a day, over a period of 2 weeks); Group 3: fructose-induced diabetic rats; and Group 4: fructose-induced diabetic rats (administered MP at a dose of 100 mg/kg. bw orally twice a day, over a period of 2 weeks). It was reported that MP extracts were able to attenuate both PG levels and the observed case of hyperinsulinemia, while glycogen levels (liver and skeletal muscle) were significantly increased in the Group 4 rats. Additionally, it was reported that the extract was able to significantly reduce the body weight of the treated rats of Group 4 (Sundaresan and Radhiga 2015).

In another report, Rajendiran et al. aimed at evaluating the antidiabetic effect of the MP leaf extract in STZ-induced diabetic rats over a 30-day period. Like the Sutar-led group, the efficacy of the extract on the diabetic rats was gauged against that of the commercial drug, metformin. It was found that oral treatment of the rats, with 300 mg/kg.bw of ethanolic extract, resulted in significant decreases in their BG and HbA1c levels, when compared to metformin. Similar to the findings of Sudaresan et al., the study group also reported a decrease in the body weight of their MP-treated subjects (Rajendiran et al. 2017).

Apart from the leaves, seeds have also been reported to possess antidiabetic properties. Kumar, Sunday, and Obuotor investigated the potential antidiabetic and antioxidant effects of MP seeds in STZ-induced diabetic Wistar rats. They used the ethanolic extract from the seeds and further fractionated it using three solvents: butanol, ethyl acetate, and hexane. The group reported that both the ethanolic extract and the associated extract fractions caused a significant reduction in the FBG levels, with the ethyl acetate fraction having the maximum effect. Furthermore, they found that all extracts were able to significantly increase serum insulin (Kumar, Sunday and Obuotor 2019).

3.38 *MOMORDICA CHARANTIA* LINN. *(MC)*

Common names: Bitter Melon, Bitter Gourd, Bitter Squash, Balsam-pear, Karela, Carailla, Carille Cerasee, Coraillie (Figure 3.66).

FIGURE 3.66 Illustration of *Momordica charantia*.

M. charantia is a slender-stemmed tendril climber of the Cucurbitaceae family. Its older stems are often flattened and fluted to 6 m or longer, and its fruits are 8–15 cm long with a distinctly bumpy exterior (I. A. Ross 2003). Originally found only in the tropics of the Old World, the plant now grows in the tropical regions of the Americas and the Caribbean (Dey, Attele and Yuan 2002, I. A. Ross 2003). MC is not only considered a food source but is also used traditionally as a medicine in developing countries due to its therapeutic properties, including anticancer, anti-inflammation, antivirus, cholesterol-lowering, and antidiabetic action. The antidiabetic effects of the MC fruit, as well as the hypoglycaemic effects of the entire plant, have been widely investigated in both experimental and clinical studies.

The main constituents of MC that have been proposed to be responsible for its BG-lowering capabilities are triterpenes, proteins steroids, alkaloids, lipids, and inorganic and phenolic compounds. However, the most efficacious compounds from its extracts were found to be charantin (Figure 3.67a), vicine (Figure 3.67b) (Joseph and Jini 2013), polypeptide-p (Dey, Attele and Yuan 2002, Joseph and Jini 2013), and the non-protein, lectin. It was further noted that MC contains other biologically active compounds such as momordicin I (Figure 3.67c), momordicin II (Figure 3.67d), cucurbitacin B (Figure 3.67e) (Kim et al. 2018, Joseph and Jini 2013), glycosides (momordin, charantin, charantosides, and goyaglycosides), terpenoid compounds (momordicinin, momordicilin, and momordol), and the cytotoxic protein (momorcharin) which also contribute to its antidiabetic activity (Khan et al. 2014).

Sarkar et al. investigated the hypoglycaemic effect of alcoholic fruit extracts (500 mg/kg.bw) from the MC fruit in both healthy and STZ-induced diabetic rats. It was found that the extract depressed the PG levels in non-diabetic subjects by 10%–16%

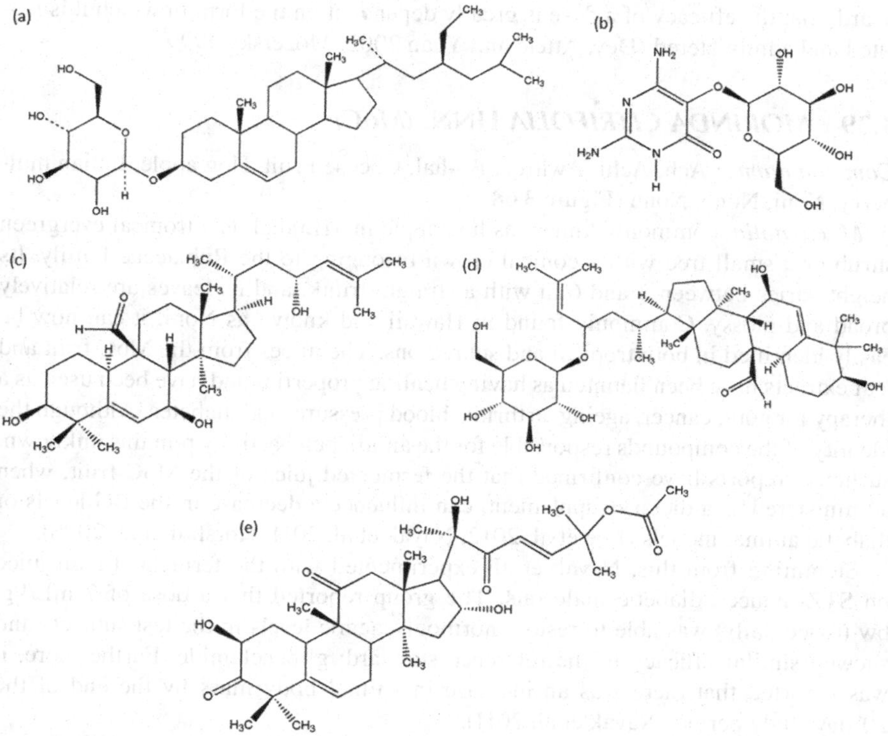

FIGURE 3.67 Structural formulae of isolated compounds from *Momordica charantia*: (a) charantin (β-sitosteryl glucoside), (b) vicine, (c) momordicin I, (d) momordicin II, and (e) cucurbatacin B.

at 1 hour and a further 6% at 2 hours. In comparison, their control standard: tolbutamide (100 mg/kg.bw), caused a 40%–44% reduction in PG at both time intervals and under similar conditions. When compared to the STZ-induced diabetic rats, MC reduced PG levels by 26% at 3.5 hours in comparison to the 40%–50% reduction of PG levels by metformin under similar conditions (Sarkar, Pranava and Marita 1996).

In another study, Virdi et al. compared the results of three types of extracts from MC using methanol, chloroform, and water. The most successful of the three was reported to be the dehydrated aqueous extract from fresh unripe whole fruits, at a dose of 20 mg/kg.bw. This dose was able to reduce the FBG by 48%, which was comparable to that of glibenclamide. A point to note is that at higher doses, the MC was deemed to be not only ineffective but also toxic! (Virdi et al. 2003)

Yet, despite the reported shortcomings, researchers such as Srivastava et al. still believe in the therapeutic use of the MC extracts due to their modest ameliorative effects. In support of this, their studies reported the effectiveness of the aqueous extract of the MC fruit versus the powdered version of the dried fruit (Srivastava, Venkatakrishna-Bhatt et al. 1993). This underscored a very valid point across the board, that the efficacy of a dose is greatly dependent on the form in which it is isolated and administered (Dey, Attele and Yuan 2002, Mozersky 1999).

3.39 *MORINDA CITRIFOLIA* LINN. *(MoC)*

Common names: Ach, Achi, Awltree, Bo-aal, Cheese Fruit, Hog apple, Indian mulberry, Noni, Nono, Nonu (Figure 3.68)

M. citrifolia, commonly known as hog apple in Trinidad, is a tropical evergreen shrub or a small tree with a conical crown belonging to the Rubiaceae family. Its height varies between 3 and 6 m with a straight trunk, and its leaves are relatively broad and glossy. Commonly found in Hawaii and known as Noni, it can now be easily identified in both tropical and subregions. The juices from the MoC fruit and leaf extracts have been flaunted as having healing properties and have been used as a therapy for gout, cancer, ageing, arthritis, blood pressure, and diabetes. Although the identity of the compounds responsible for the antidiabetic activity remains unknown, sufficient reports have confirmed that the fermented juice of the MoC fruit, when administered as a dietary supplement, can influence a decrease in the BG levels of diabetic animal models (Lee et al. 2012, Nayak et al. 2011, Horsfall et al. 2008).

Stemming from this, Nayak et al. experimented with the fermented fruit juice on STZ-induced diabetic male rats. The group reported that a dose of 2 mL/kg. bw (twice daily) was able to restore normoglycaemic levels in the test subjects and showed similar efficacy to the reference standard: glibenclamide. Furthermore, it was reported that there was an increase in animal body mass by the end of the 20-day study period (Nayak et al. 2011).

In another study, Horsfall et al. investigated the synergistic effects of Noni juice with insulin (humilin 70/30), using alloxan-induced diabetic rats. Interestingly, the results of their experiments showed that at the end of the 4-week study period, the mean FBG level of the rats that received the combination therapy of juice and insulin was lower than that of either therapies, when administered by itself (Horsfall et al. 2008).

FIGURE 3.68 Illustration of *Morinda citrifolia*.

3.40 *MORINGA OLEIFERA* LINN. *(MO)*

Common names: Drumstick tree, Horseradish tree, Moringa, Mother's best friend, Saijan, West Indian ben (Figure 3.69)

M. oleifera is a fast-growing, drought-resistant tree of the family: Moringaceae that was originally native to the Himalayas, but is currently cultivated in many tropical and subtropical regions around the world. Having such a rich history of its use in Ayurvedic medicine, all parts of the plant (the leaves, fruits, flowers, bark, and roots) are now being thoroughly explored using modern techniques to confirm their therapeutic applications to various diseases (Ahmad, Khan and Blundell 2019, Vargas-Sánchez, Garay-Jaramillo and González-Reyes 2019, Tang et al. 2017, Jaiswal et al. 2009, Latif et al. 2014).

Regarding its applications to DM therapy, several mechanisms have been proposed that address the possible and plausible interactions of the bioactive isolates. These are represented by the relationship chart in Figure 3.70 (Ahmad, Khan and Blundell 2019). However, as new isolates are reported, we can expect that further mechanism bubbles will be added in the future.

The leaves of MO were found to be rich in proteins, vitamins, minerals, β-carotene, and other biologically active compounds that include dietary fibre, flavonoids, phenolic acids, alkaloids, carotenoids, isothiocyanates, glucosinolates, tannins, saponins, oxalates, and phytates. From the list of bioactive compounds isolated, 4-hydroxyphenylacetonitrite (Figure 3.71a), fluoropyrazine (Figure 3.71b), methyl-4-hydroxybenzoate (Figure 3.71c), and vanillin (Figure 3.71d) have been isolated from its leaves and have been shown to influence both a significant glucose-dependent insulin release (at a stimulatory glucose concentration of 16.7 mM),

FIGURE 3.69 Illustration of *Moringa oleifera*.

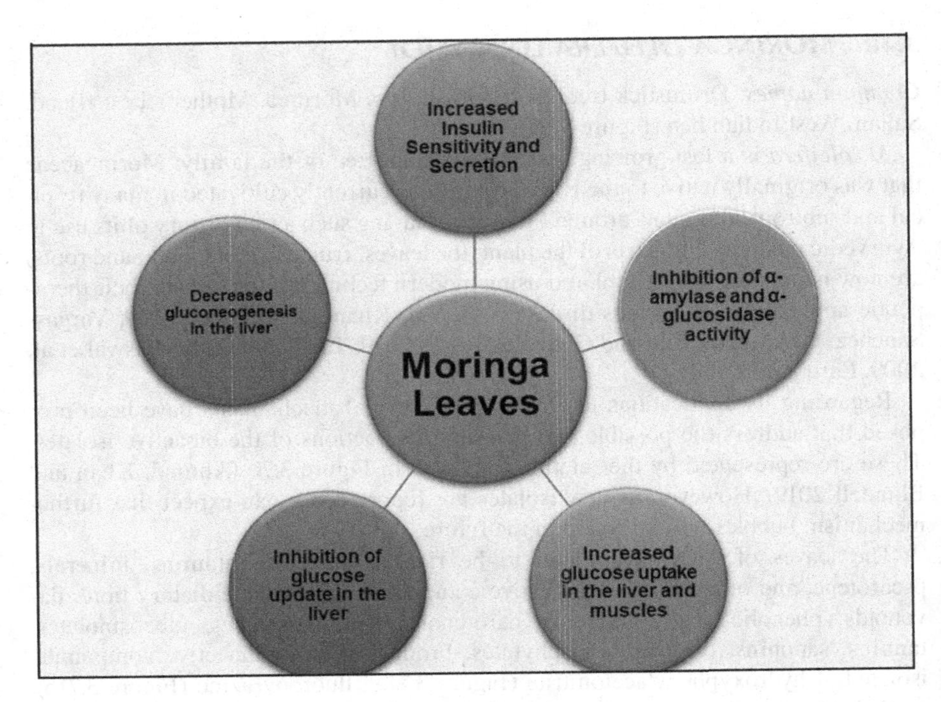

FIGURE 3.70 Summary of the possible mechanisms of hypoglycaemic action of MO leaf extracts.

and a dose-dependent release of insulin (at a dose of 200 μM) (Ahmad, Khan and Blundell 2019).

Further to the aforementioned isolates, other researchers have reported the discovery of several polyphenols in MO extracts. Among the most important are the flavonoids: kaempferol (Figure 3.25f) and quercetin (Figure 3.25g), and the phenolic acids: chlorogenic acid (Figure 3.25c) and caffeoylquinic acid (Figure 3.71e), which exhibit antihyperglycaemic properties by acting as competitive inhibitors of the sodium-glucose linked transporter type 1 (SGLT1) in the mucosa of the small intestine. Interestingly, their mechanisms of action seems to mimic those of both the allopathic SLGTi and the AG inhibitors (see Chapter 1.5 and 1.7) (Vargas-Sánchez, Garay-Jaramillo and González-Reyes 2019).

The potency of yet another plant isolate was demonstrated by the Waterman group, who showed that the oral administration of a 5% moringa concentrate (containing an estimated 66 mg/kg.bw of the biologically active moringa isothiocyanate [MIC]: moringin Figure 3.71f]) to obese-C57BL/6L mice led to a significant reduction in the accumulated fat mass and an improvement in glucose tolerance and insulin signalling (Waterman et al. 2015).

Although the following final report may be argued to not belong here because of the nature of the administration, we believed that it was still worthy of mention, due to the link between the active compound and its botanical source. Paula et al.

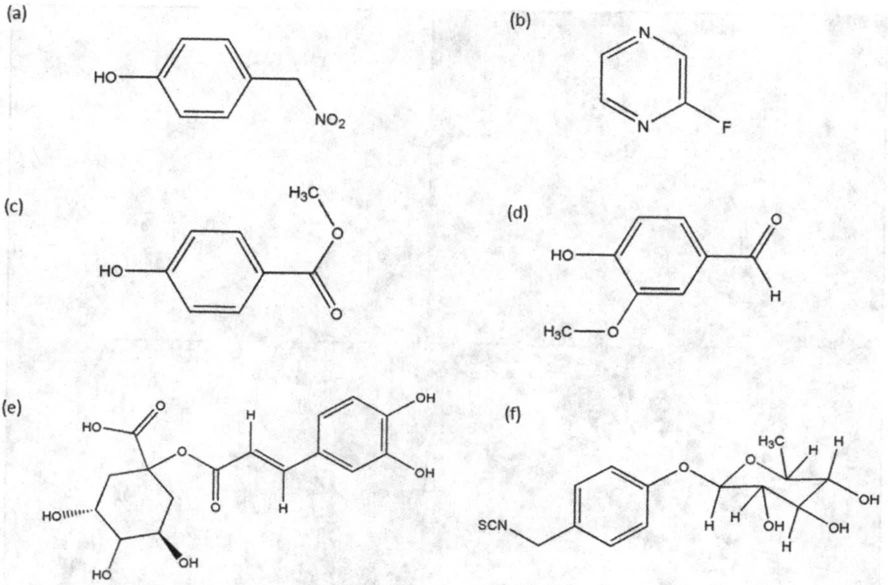

FIGURE 3.71 Structural formulae of isolated compounds from *Moringa oleifera*: (a) 4-hydroxyphenylacetonitrite, (b) fluoropyrazine, (c) methyl-4-hydroxybenzoate, (d) vanillin, (e) caffeoylquinic acid, and (f) moringin (4-(α-L-rhamnosyloxy)benzyl isothiocyanate).

confirmed the antidiabetic activity of the leaf isolate (*M. oleifera* leaf protein isolate [MO-LPI]) by giving a single intraperitoneal dose of the compound (at 300 or 500 mg/kg.bw) to diabetic mice. The group reported that a more pronounced effect was observed when the higher dose of 500 mg/kg.bw was used, resulting in drops in the BG levels by 34.3%, 60.9%, and 66.4% after 1, 3, and 5 hours, respectively. The sustained BG-lowering potential of MO-LPI was also evidenced when the intraperitoneal administration of the higher dose for 1 week resulted in a reduction in the mice's BG levels by 56.2%. As in the case with many other plant extracts, with such promising results, one may expect that concerns would be raised regarding the associated compound toxicity. However, it was indeed eye-catching to learn that there were no reported adverse effects, even at an extreme dose of 2500 mg/kg.bw (Paula et al. 2017).

3.41 *MURRAYA KOENIGII* LINN. *(MK)*

Common names: Caripeelay, Carypoulay, Curry leaf, Curry patta, Kaddi patta (Figure 3.72)

M. koenigii is a small, tropical to subtropical tree or shrub with an average height of 4–6 m and a trunk up to 40 cm in diameter. The aromatic leaves are pinnate, with 11–21 leaflets, each leaflet 2–4 cm long and 1–2 cm broad. The plant produces small white flowers that can self-pollinate to produce small shiny-black drupes containing

FIGURE 3.72 Illustration of *Murraya koenigii*.

a single, large viable seed, and berry pulp, which is edible, with a sweet flavour (Wikipedia 2021). This tree is native to moist forests in India and Sri Lanka, but was introduced to the West Indies during the colonization period.

The MK plant is highly reputed for its stimulant, stomachic, antidysenteric (Wang et al. 2003), and antidiabetic properties (Husna et al. 2018, Kesari et al. 2007, Vinuthan et al. 2004), which are owed to the wide variety of biological substances, such alkaloids, flavonoids, terpenoids, tannins, glycosides, and phenolics that are contained within. Preliminary phytochemical screening by Husna et al. showed the presence of the bioactive compounds: 1,8-cineol (Figure 3.73a), β-caryophyllen (Figure 3.73b), hexadecen-1-ol (Figure 3.73c), α-matrine (Figure 3.73d), benzo[a]naphtacene (Figure 3.73e), γ-sitosterol (Figure 3.73f), and vitamin E (Figure 3.73g) (Husna et al. 2018). Furthermore, 1-formyl-3-methoxy-6-methylcarbazole (Figure 3.73h) and 6,7-dimethoxy-1-hydroxy-3methylcarbazole (Figure 3.73i) were later isolated by the Chowdhury group (Wang et al. 2003).

Yadav et al. investigated the efficacy of feeding MK leaves as a dietary constituent to control hyperglycaemia in alloxan-induced rats. It was found that the supplement did not show any profound hypoglycaemic effects in the 7-day study. However, the potential therapeutic effects were not dismissed due to the probability that the period of time for the effect to be noticeable may have been longer than the study period. That aside, the group also noticed an unusual outcome from the MK

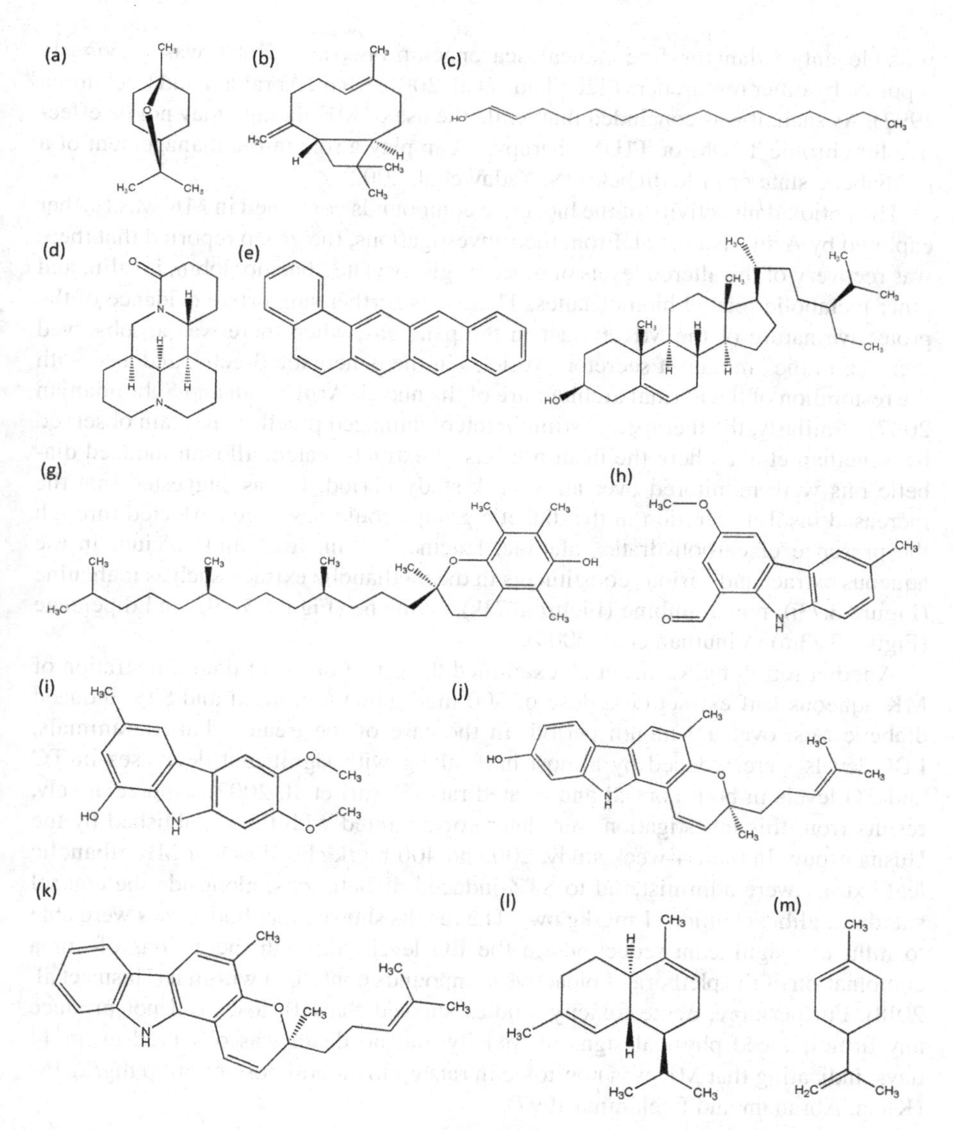

FIGURE 3.73 Structural formulae of *Murraya koenigii*: (a) 1,8-cineol, (b) β-caryophyllen, (c) hexadecen-1-ol, (d) α-matrine, (e) benzo[a]naphtacene, (f) γ-sitosterol, (g) vitamin E, (h) 1-formyl-3-methoxy-6-methylcarbazole, (i) 6,7-dimethoxy-1-hydroxy-3-methylcarbazole, (j) mahanine, (k) mahanimbine, (l) cadinene, and (m) dipentene.

supplementation, which they coined an "anti-alloxan" effect. Alloxan and STZ are chemicals known to produce hyperglycaemia through selective cytotoxic effects on the pancreatic β-cells; one of the intracellular phenomena for their cytotoxicity was discovered to be the generation of free radicals that target the pancreas. It was suggested in the past, that MK possibly prevented the destruction of β-cells through possible antioxidant or free radical scavenger properties, which was previously reported by other researchers (Tachibana et al. 2001, Khan, Abraham and Leelamma 1997). As such, it was concluded that while the use of MK therapy may not be effective for chronic T2DM or T1DM therapy, it can play a role in the management of a prediabetic state or mild diabetes (S. Yadav et al. 2002).

The antioxidant activity of the bioactive compounds contained in MK was further explored by Arulselvan et al. From their investigations, the group reported that there was recovery of the altered levels of glucose, glycosylated haemoglobin, insulin, and other metabolic-related biomolecules. There was further supportive evidence of the protective nature of the MK extract in the pancreas, when there was an observed increase in the amount of secretory vesicles in the pancreatic β-cells, and also with the restoration of the normal architecture of the nuclei (Arulselvan and Subramanian 2007). Similarly, the therapeutic stimulation of damaged β-cells was again observed by Vinuthan et al., where the insulin levels of extract-treated, alloxan-induced diabetic rats were monitored over an 8-week study period. It was suggested that the increased insulin secretion in the diabetic groups could have been effected through the presence of carbohydrates, alanine, leucine, niacin, iron, and calcium in the aqueous extract and various constituents in the methanolic extract, such as mahanine (Figure 3.73j), mahanimbine (Figure 3.73k), cadinene (Figure 3.73l), and dipentene (Figure 3.73m) (Vinuthan et al. 2004).

Another report by Kesari et al. examined the effect of the oral administration of MK aqueous leaf extract (at a dose of 300 mg/kg.bw) to normal and STZ-induced diabetic rats, over a 1-month period. In the case of the treated diabetic animals, FBG levels were reduced by almost half, along with significant decreases in TC and TG levels in both normal and treated rats (Kesari et al. 2007). Appreciatively, results from this investigation were later corroborated with those published by the Husna group. In their 4-week study, 200 and 400 mg/kg.bw doses of MK ethanolic leaf extract were administered to STZ-induced diabetic rats, alongside the control standard, glibenclamide (1 mg/kg.bw). The results showed that both doses were able to influence significant reductions in the BG levels, through the actions of-, or a combination of the plethora of bioactive compounds contained within it (Husna et al. 2018). Furthermore, acute toxicity studies showed that all doses did not produce any drug-induced physical signs of toxicity and no death was observed up to 14 days, indicating that MK was non-toxic in rats up to an oral dose of 5000 mg/kg.bw (Khan, Abraham and Leelamma 1997).

3.42 *NEUROLAENA LOBATA* LINN. *(NL)*

Common names: Herbe á pic, Jackass Bitters, Zeb-a-pique (Figure 3.74)

N. lobata is a herb or bush (Andrade-Cetto, Cruz et al. 2019) of the Asteraceae or Compositae family that inhabits warm and semi-warm climates. It stands erect with stems up to 3 m high (Cruz and Andrade-Cetto 2015, Gupta et al. 1984, Morales et al. 2001), and with long, slender leaves when young, but typically having three points when mature. The leaves of the NL plant are considered to exhibit hypoglycaemic properties (Gupta et al. 1984, Morales et al. 2001) and to date, although several

FIGURE 3.74 Illustration of *Neurolaena lobata*.

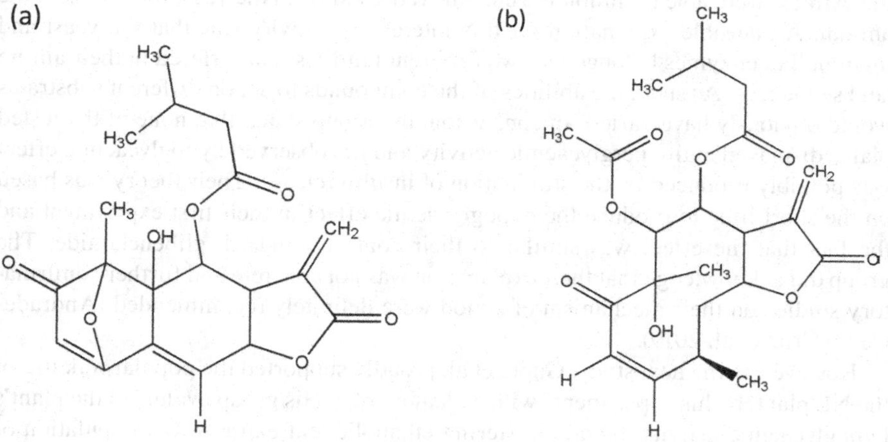

FIGURE 3.75 Structural formulae of *Neurolaena lobata*: (a) lobatin-B and (b) neurolenin-B.

isolates have been proposed to be responsible for this desirable action, there seems to be some agreement that the compounds, lobatin-B (Figure 3.75a) and neurolenin-B (Figure 3.75b), are the most efficacious.

While there have been no reported cases of clinical studies performed using the NL plant, a study by Andrews et al. looked at its history of use, along with other herbal remedies in the rural areas of Guatemala. It was heartening to learn of the objective of the authors, such that although traditional healers were not the main source of diabetic treatment, they recognized that many patients still relied on medicinal plants to control their blood sugar, due to a lack of access to allopathic medicines. As such, documenting the medicinal plants used to treat diabetes was essential to improving diabetic care in the area. Since a large proportion of people with diabetes in these communities use medicinal plants, it was important for healthcare providers, including health promoters and traditional healers, to understand their proper use and potential interactions (Andrews, Wyne and Svenson 2018).

In this first instance, a study was carried out to compare the hypoglycaemic effect of four plants (*Hamelia patens* Linn. [HP], *N. lobata*, *Solanum americanum* Linn. [SA], and the cortex of *Croton guatemalensis* Linn. [CG]), all of which were native to Guatemala. The study group's first experiment aimed at gauging the plants' relative antidiabetic activities under normal conditions (without a sugar load) and then under a maltose- and sucrose-loaded condition. In agreement with the traditional usage of the plants, in normal conditions, the extracts produced a statistically significant hypoglycaemic effect, similar to the control drug, glibenclamide. However, when the sugar was sucrose, both CG and HP produced a statistically significant effect at 30 minutes, compared to the control group, while SA did so at 60 minutes, and NL was slowest in activity (at 90 minutes) (Andrade-Cetto, Cruz et al. 2019).

In their next study, the inhibitory potential of α-glucosidase enzymes, which were taken from either yeast or rat intestines, was investigated. These assays showed that the extracts were able to inhibit the enzyme retrieved from the yeast but not from the animal. A plausible explanation for this interesting activity was that the yeast and mammalian enzymes belonged to two different families that differed in their amino acid sequences. As such, the abilities of the compounds to act on different substrates would obviously have varied. In conclusion, the group stated that none of the tested plants displayed antihyperglycaemic activity and the observed hypoglycaemic effect was possibly produced by the stimulation of insulin release. Their theory was based on the short time to produce the hypoglycaemic effect in their first experiment and the fact that the effect was similar to their control standard, glibenclamide. The group did acknowledge that their explanation was not concrete and further confirmatory studies on their mechanism of action were definitely recommended (Andrade-Cetto, Cruz et al. 2019).

However, in the next study, Gupta et al. proudly supported the popular folk use of the NL plant, by his experiments with its leaf extract. His group evaluated the plant's hypoglycaemic activity, by administering ethanolic leaf extracts to a population of 12 alloxan-induced diabetic mice. The results of this study showed that the extracts (at doses of 250 and 500 mg/kg.bw) were able to lower elevated BG levels in the tested nomoglycaemic and hyperglycaemic mice, respectively, within a space of 4 hours (Gupta et al. 1984)

Apart from its direct antidiabetic potential, the plant has also been proposed to have ameliorative effects in diabetes-related illnesses. Regarding this, Morales et al. aimed at evaluating the influence of several popular native plants of Guatemala on the occurrence of toxic events in the central nervous system of Swiss albino mice. Aqueous and organic extracts from the areal parts of *Tridax procumbens* Linn. (TP), *N. lobata* (NL), *Byrsonima crassifolia* Linn. (BC), and *Gliricidia sepium* Linn. (GS), and of the root and leaves of *Petieria alliacea* Linn. (PA) were used. Their report confirmed that the aqueous leaf-extract of NL was able to modestly decrease the spontaneous motor activity and muscle tone and also mice-exhibited diuresis. However, the authors concluded that in relation to the other plants screened, the NL plant showed relatively little or no effect on behavioural, neurological, and autonomic signs (Morales et al. 2001).

3.43 *OCIMUM GRATISSIMUM* LINN. *(OG)* AND *OCIMUM TENUIFLORUM* LINN. *(OT)*

Common names: *Ocimum gratissimum*: Clove basil, African basil, Wild basil (Figure 3.76a);
Ocimum tenuiflorum: East Indian basil, Holy basil, Krishna Toolsie, Tulasi, Tulsi (Figure 3.76b)

Among all the medicinal plants, those under the genus of *Ocimum* (belonging to family Lamiaceae) have strong therapeutic potentials. These herbs are commonly found in the tropical regions of Asia, Africa, and all of the Americas. Several *Ocimum* species exhibit strong antihyperglycaemic effects (Antora and Salleh 2017);

FIGURE 3.76 Illustrations of *Ocimum* species: (a) *Ocimum gratissimum* and (b) *Ocimum tenuiflorum*.

however, two species (*O. gratissimum* and *O. tenuiflorum*) are of particular interest due to their relative ubiquity in the North American and Caribbean (NAC) region.

The extracted essential oil from the OG plant has shown high concentrations of eugenol (1-hydroxy-2-methoxy-4-allybenzene) (Figure 3.77a) and chicoric acid (Figure 3.77b), which have been attributed to its prowess in the attenuation of hyperglycaemia (Antora and Salleh 2017).

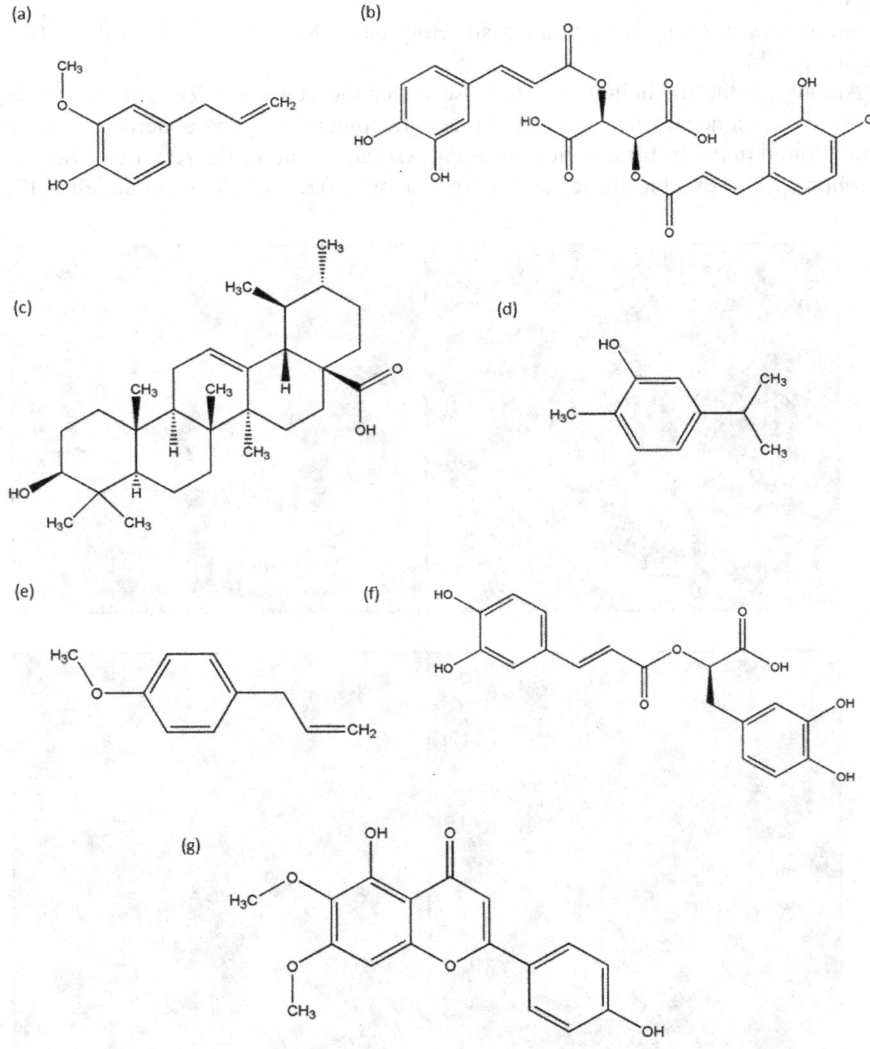

FIGURE 3.77 Structural formulae of isolated compounds from *Ocimum gratissimum* and *Ocimum tenuiflorum*: (a) eugenol, (b) chicoric acid, (c) ursolic acid, (d) carvacrol, (e) estragole, (f) rosmarinic acid, and (g) cirsimaritin.

Reports have shown that the aqueous leaf extract of the OG plant can drastically reduce postprandial hyperglycaemia in T2DM model rats, but without the risk of hypoglycaemia (Oguanobi, Chijioke and Ghasi 2012). Supporting studies by the Ekaiko group, have also revealed that the elevated BG levels in diabetic rats were reduced, using doses of 200 and 400 mg/kg.bw of aqueous leaf extracts (Ekaiko et al. 2016).

Another investigating team showed that intraperitoneal treatment with the methanolic extract of the OG leaves (400 mg/kg.bw) reduced the BG levels in both normal and diabetic rats (Aguiyi et al. 2000). Furthermore, the team led by Okoduwa demonstrated that the daily administration of *n*-hexane, chloroform, ethyl acetate, *n*-butanol, and the aqueous fractions of the OG leaf resulted in lowered BG levels in a T2DM animal model (Okoduwa et al. 2017). It was also confirmed that at equal doses, the aqueous extracts of OG leaves, led to greater drops in BG levels than the ethanolic extracts (Antora and Salleh 2017).

We have also stumbled upon the suggestion that the OG leaf extracts can stimulate surviving β-cells to release more insulin in T2DM subjects (Oguanobi, Chijioke and Ghasi 2012, Okoduwa et al. 2017) and that the hypoglycaemic efficacy of OG was higher than that of insulin! (Okon and Umoren 2017).

Regarding OT (formerly known as *Ocimum sanctum* Linn.), this species is particularly sacred in India, and its fresh leaves are commonly used in the treatment of coughs, colds, abdominal pain, skin diseases, arthritis, painful eye diseases, measles, and diarrhoea. Furthermore, the preclinical evaluation of various extracts of different parts of OT showed antifertility, anticancer, antifungal, cardioprotective, hepatoprotective, and antidiabetic actions.

In this first instance, Parasuraman et al. reported that the hydro-alcoholic extract of OT exhibited significant antidiabetic activity at doses of 250 and 500 mg/kg.bw, in STZ- and NIC--induced diabetic rats, and was comparable to that of glibenclamide. Some phytoconstituents identified as influential to its efficacy were eugenol (Figure 3.77a), ursolic acid (Figure 3.77c), carvacrol (Figure 3.77d), linalool (Figure 3.45b), estragole (Figure 3.77e), rosmarinic acid (Figure 3.77f), apigenin (Figure 3.14b), and cirsimaritin (Figure 3.77g) (Parasuraman et al. 2015).

Another study that focused on the phytochemical qualitative assay of aqueous crude extract and aqueous fractions reported that the active methanolic crude extract and its active fractions (ethyl acetate and butanol) showed the presence of polyphenolic active constituents (3,4-dimethoxycinnamic acid, caffeic acid, diosmetin, luteolin, kaempferol, and genistein), which are all well known to exhibit antidiabetic activity (Mousavi, Salleh and Murugaiyah 2018).

It was unusual to also find a study that was aimed at evaluating the effect of the OT aqueous leaf extract in hyperglycaemic tilapia (*Oreochromis niloticus* Linn.). Interestingly, the group also confirmed the extract's antihyperglycaemic activity in this form of animal! (Arenal et al. 2012).

Regarding toxicity, one study showed that a dose of 300 mg/kg.bw of OG leaf extract (both aqueous and ethanolic) appeared to have no adverse effects on the organs of the test subjects (Oguanobi, Chijioke and Ghasl et al. 2019). It was also indicated that a dose of 350 mg/kg.bw of combined aqueous leaf- extracts of OG and

G. latifolium Linn. showed no toxicity (Eyo and Chukwu 2016) with the lethal dose estimated as ≥5000 mg/kg.bw (Eyo and Chukwu 2016, Lawal et al. 2019). Adding to this, another acute toxicity study reported that a single oral administration of combined extracts of OG and *V. amygdalina* Linn. (at doses from 10 to 5000 mg/kg.bw) did not produce any apparent toxic symptoms or mortality after a 24-hour period (Okoduwa et al. 2017, Abdulazeez et al. 2013). Despite these promising revelations, caution has still be placed on its use, as studies have suggested that the constituent compounds in OG may worsen diabetes-induced renal injury (Onaolapo, Onaolapo and Adewole 2012).

With respect to OT, a study carried out by Mousavi et al. concluded that based on the histological analysis of tissues, there were no adverse changes in the liver, kidneys, and pancreas of the test subjects in comparison to the control group (Mousavi, Salleh and Murugaiyah 2018). Similarly, the previously discussed study by Parasuraman et al., using the hydro-alcoholic extract of OT, did not show any harmful effects and also showed no mortality up to 2000 mg/kg.bw, when given as a single oral administration (Parasuraman et al. 2015).

3.44 *PANAX GINSENG* LINN. *(PG)* AND *PANAX QUINQUEFOLIUS* LINN. *(PQ)*

Common names: *Panax ginseng*: Asian Ginseng, Chinese Ginseng, Guigai, Jiln Ginseng, Korean Ginseng, Ninjin, Oriental Ginseng, Panax schinseng, Red Ginseng (Figure 3.78); *Panax quinquefolius*: American ginseng, Canadian ginseng, Ginseng

Of the different types of ginseng, *P. ginseng* and *P. quinquefolius* are the most commonly used for medicinal purposes. The perennial herb PG is considered native to countries in the Oriental region, such as Korea, Siberia, and China, and has also been found growing wild and cultivated in North America. It has fleshy, pale-yellow, multi-branched roots with an average stem height of 0.6–0.8 m (Schulz et al. 2004).

Most of the active ingredients found in the ginseng species are ginsenosides, polysaccharides, peptides, polyacetylenic alcohols, phenolic compounds, and fatty acids (Dey, Attele and Yuan 2002, Kim et al. 2016). However, the ginsenosides (bearing protopanaxadiol or protopanaxatriol groups) (Figure 3.79) have been found to be the most pharmacologically active components (Dey, Attele and Yuan 2002, Lim et al. 2009) and this has warranted their therapeutic use in the treatment of T2DM (Rahimi 2015).

Extracts of both ginseng species show antihyperglycaemic activity (Lim et al. 2009) and are more specifically associated with increased PPAR-γ expression and AMP-activated protein kinase phosphorylation in the liver and muscle (Lim et al. 2009, Rahimi 2015). Additional biochemical interactions, as stated by Rosalie and Ekpe, include a decreased carbohydrate absorption rate into the hepatic portal circulatory system, increased glucose transport and uptake (mediated by nitric oxide), increased glycogen storage, and the modulation of insulin secretion (Dey, Attele and Yuan 2002, Vuksan et al. 2000, Rosalie and Ekpe 2016, Verma et al. 2018). Furthermore, Lim et al. suggested that ginsam (a vinegar extract from PG) has

FIGURE 3.78 Illustrations of *Panax* species: (a) *Panax ginseng* and (b) *Panax quinquefolius.*

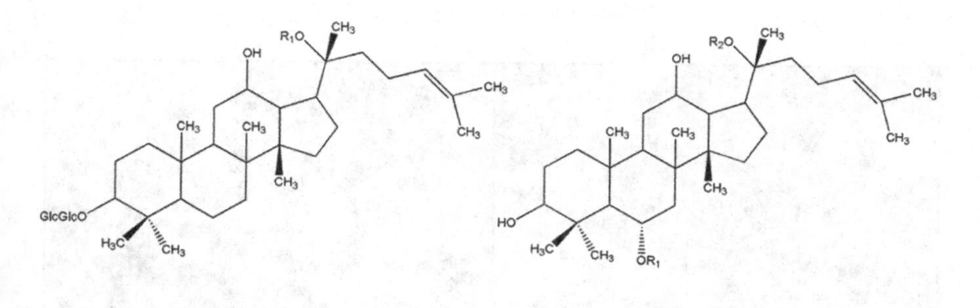

Ginsenosides	R_1
Rb1	Glc-^{6}Glc
Rb2	Ara-^{6}Glc
Rc	Ara-^{6}Glc
Rd	Glc

Ginsenosides	R_1	R_2
Re	Glc-^{2}Rha	Glc
Rf	Glc-^{2}Glc	H
Rg1	Glc	Glc
Rg2	Glc-^{2}Rha	H

FIGURE 3.79 Structural formulae of common *Panax ginseng* and *Panax quinquefolius* ginsenosides.

distinct beneficial effects on glucose metabolism (Lim et al. 2009). As such, it was deemed possible to be an adjuvant therapy for treating diabetic patients with insulin resistance (Rahimi 2015).

Sotaniemi et al. demonstrated a reduction in both FBG and A1C (HbA1c) levels in persons with T2DM when treated with a 100 and 200 mg dose of the extracts, respectively (Dey, Attele and Yuan 2002, Vuksan et al. 2000, Sotaniemi, Haapakoski and Rautio 1995). However, Schulz et al. suggested that a more reliable therapeutic effect could be achieved with a dosage of 1–2 g (crude) or 200–600 mg of extract (Dey, Attele and Yuan 2002, Schulz et al. 2004).

It should be noted as a precautionary measure, that consistent use of ginseng can consequently affect the central nervous system (Dey, Attele and Yuan 2002, Bailey and Day 1989), therefore it is important that the dosage should be monitored and reduced if necessary. As such, some researchers strongly recommend that there should be a 3-month limit on the period of treatment (Dey, Attele and Yuan 2002).

3.45 *PARTHENIUM HYSTEROPHORUS* LINN. *(PaH)*

Common names: Santa-Maria, Santa Maria feverfew, Whitetop weed, Famine weed, Whitehead broom (Figure 3.80)

Belonging to the Asteraceae family, *P. hysterophorus* is a short flowering plant that is mainly located in Asia, Africa, the Americas, and the Caribbean. It is easily identified as ground weeds found along roadsides and in backyards. In traditional medicine, reports have revealed that PaH is used in a wide range of therapies due to its antiviral, antifungal, antibacterial, anti-helminthic, anti-inflammatory, and anti-diabetic properties (Joshi et al. 2016, Mew et al. 1982).

FIGURE 3.80 Illustration of *Parthenium hysterophorus*.

Patel et al. investigated the hypoglycaemic effects of PaH in alloxan-induced diabetic rats with BG levels in the range of 280–310 mg/dL. The rats were divided into three groups and administered the following solutions via oral gavage: control (only solvent); positive control (0.5 mg/kg.bw glibenclamide + solvent); and test (100 mg/kg.bw PaH aqueous flower extract + solvent). In relation to the positive control, the results showed a significant reduction in the BG levels after 2 hours (<240 mg/dL) and also a significant decrease in the FBG levels in the treated rats (Patel et al. 2008).

Arya et al. also investigated the hypoglycaemic effects of PaH in diabetic rats, using fresh-leaf extract (Arya et al. 2012). However, when compared to the previously discussed study group, we found that the attenuation of the BG levels was slightly less appreciable (Arya et al. 2012, Sahrawat et al. 2018). As promising as the results from both studies were, concurrently run toxicity evaluations have revealed a precautionary toxic nature of PaH (Arya et al. 2012). Supporting this, additional studies have shown that regular exposure to the chemicals contained within the extracts may lead to dermatitis and respiratory problems (Khaket et al. 2015, Towers and Subba Rao 1992, McFadyen 1992).

3.46 *PEPEROMIA PELLUCIDA* LINN. *(PP)*

Common names: Pepper elder, Silverbush, Rat-ear, Shining bush plant, Man to Man (Figure 3.81)

P. pellucida is a member of the Piperaceae family and is an annual common weed that is delicate, fleshy, and glabrous. It usually grows to a length of 15–45 cm with translucent, erect, pale-green stems with internodes usually 3–8 cm long. PP is native to tropical North and South America (Amarathunga and Kankanamge 2017, Raghavendra and Prashith Kekuda 2018), but is now widely distributed throughout the tropics (Amarathunga and Kankanamge 2017, N et al. 2018). Several studies (using alloxan-induced diabetic rodent models) have been published that confirm the

FIGURE 3.81 Illustration of *Peperomia pellucida*.

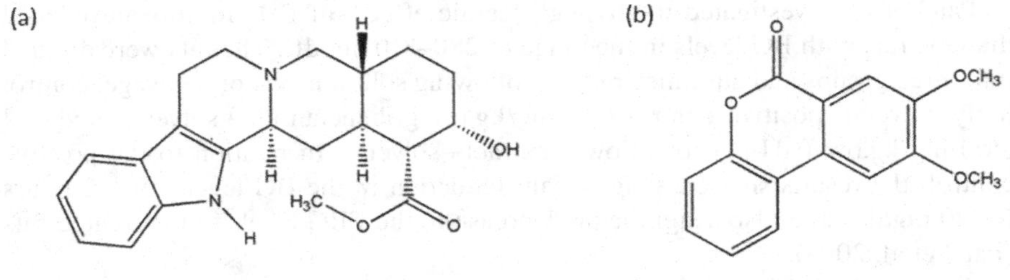

(a) (b)

FIGURE 3.82 Structural formulae of isolated compounds from *Peperomia pellucida*: (a) yohimbine and (b) 8,9-dimethoxyellagic acid.

plant's hypoglycaemic activity (Hamzah et al. 2012, N et al. 2018, Kanedi, Sutyarso et al. 2019, Kanedi, Sutyarso et al. 2019).

An interesting report by Hamzah et al. demonstrated that feeding diabetic rats with diets, enriched with 10% w/w and 20% w/w dried-leaf powder, resulted in a reduction in the BG levels as well as the TC, TG, and LDL levels, compared to their untreated diabetic counterparts. Furthermore, treatment with glibenclamide and PP (10% w/w and 20% w/w) led to the increased activities of SOD, catalase activities (CAT), and glutathione (GSH), respectively (Hamzah et al. 2012).

A paper by Sheikh et al. suggested that the plant's ability to stimulate or regenerate the pancreatic β-cells was attributed to its high content of the biologically active alkaloids, saponins, and flavonoids (Figure 3.4) (Sheikh et al. 2013). Apart from this, while screening the activity of the PP extracts against the receptor protein aldose reductase, Akila et al. was able to confirm the outstanding potency of one of the plant's most active constituents, yohimbine (Figure 3.82a), against the standard compound, quercetin (Figure 3.25g) (Akhila, Aleykutty and Manju 2012).

Adding to this, Susilawati et al. further isolated the compound 8,9-dimethoxyellagic acid (Figure 3.82b) from PP and determined that a dose of 100 mg/kg.bw was able to yield significant antidiabetic effects in alloxan-induced diabetic mice (Susilawati et al. 2017).

In this final report, Oyolede *et al.* performed acute brine shrimp toxicity studies (using various extracts of the PP leaves). The group recounted that the fractions with low-polarity solvents (hexane and ethyl acetate) were found to be toxic, while the most polar fractions, which contained butanol and water, were deemed safe. In the three assays used for antioxidant screening, the extracts were most effective in the hydrogen peroxide assay: giving a percentage inhibition of over 98% scavenging activity. This high efficacy (at low concentration) supports its use in the treatment of a wide range of ailments that are associated with oxidative stress. However, the authors highly recommended that the use of such toxic chemical compounds (at high doses) should be properly monitored (Oloyede, Onocha and Olaniran 2011).

3.47 *PHYLLANTHUS AMARUS* LINN. *(PA)*

Common names: Gale of the wind, Stonebreaker, Seed-under-leaf (Figure 3.83)

Ph. amarus is a small, erect, annual herb of the family Euphorbiaceae that grows to 30–60 cm, in height. It is believed to be native to the Caribbean, but is now located

FIGURE 3.83 Illustration of *Phyllanthus amarus.*

in both tropical and subtropical parts of the world (Meena, Sharma and Rolania 2018). Clinical studies have shown that extracts of PA exhibited significant hypoglycaemic effects in animals, due to its rich content of alkaloids, tannins, saponins, anthraquinones, cardiac glycosides, and flavonoids (A. A. Adeneye 2012, Adeneye, Amole and Adeneye 2006, Joseph and Raj 2011, Moshi et al. 2001, Meena, Sharma and Rolania 2018, Adedapo, Ofuegbe and Oguntibeju 2014).

Raphael et al. reported that the methanolic extract of PA was found to have potential anti-oxidant activity, as it could inhibit lipid peroxidation and scavenge hydroxyl and superoxide radicals *in vitro*. Furthermore, 4 hours after administration, the extract (at doses of 200 and 1000 mg/kg.bw) was found to reduce the BG levels in alloxan-induced diabetic rats by 6% and 18.7%, respectively. It was further reported that the rats treated with the 1000 mg/kg.bw dose, displayed normoglycaemic BG levels on Day-18, after alloxan administration (Raphael, Sabu and Kuttan 2002).

Similar doses were used by Lawson-Evi et al. when the group investigated the antidiabetic effect of aqueous and hydro-alcoholic extracts of PA on similarly diabetized rats. However, in this study the group reported a restoration of euglycaemic levels after 15 days of administration (Lawson-Evi et al. 2011). Interestingly, we found lower but equally therapeutic doses being used by another group. In that study, the oral administration of the ethanolic leaf extract (400 mg/kg.bw) for 45 days to diabetic mice, resulted in a significant decline in the BG levels (from 310.2 to 141.0 mg/dL) and a significant recovery in body weight. There was also an observed alteration in the activity levels of the liver enzymes, whereby there was a significant decrease in glucose-6-phosphatase and fructose-1-6-disphosphatase action, and a simultaneous increase in the activity of glucokinase (Shetti, Sanakal and Kaliwal 2012).

In a clinical study, Moshi et al. reported that there was no effect on non-insulin-dependent diabetes mellitus (NIDDM) patients who underwent PA extract therapy. In their experiment, 21 NIDDM patients (who were currently undergoing allopathic treatment) were asked to switch to PA herbal therapy for 1 week, following a 1 week "wash-out" period (a period where no form of T2DM therapy would be taken). It was noted that after 1 week of herbal treatment, no hypoglycaemic activity was observed.

As such, the researchers concluded that a 1-week treatment with the aqueous PA extract was incapable of lowering both the FBG and PBG levels in untreated NIDDM patients (Moshi et al. 2001). However, the results can be considered debatable, due to the relatively short study time in relation to other previously discussed study periods.

Regarding its safety in use, a dose-response study was performed by Adedapo et al., using random groups of male Wistar rats. Graded doses of the plant's extract (100, 200, 400, 800, and 1600 mg/kg.bw) were orally administered to the subjects in each of the groups, after which they were allowed free access to food and water and monitored for any signs of toxic-related events over a 48-hour period. At the end, all the rats appeared to be normal and none of them showed any visible signs of illness. Appreciatively, it was noted than even at the extreme dose of 1600 mg/kg .bw, the extract from the plant still appeared to be safe for medicinal use (Adedapo, Ofuegbe and Oguntibeju 2014).

3.48 *SCOPARIA DULCIS* LINN. *(SD)*

Common names: Licorice weed, Goatweed, Scoparia-weed, Sweet-broom (Figure 3.84)

S. dulcis is an erect, annual herb that can reach a height of 0.5 m at maturity. It belongs to the family Scrophulariaceae, and is widely distributed in tropical and subtropical regions. The plant is known to be a rich source of flavones, terpenes, and steroids and its leaves (in both fresh and wet forms) are traditionally used as remedies for DM in India (Mishra, Behera et al. 2011). However, we have noted cases in which the entire plant (inclusive of the roots) has been used (Pamunuwa, Karunaratne and Waisundara 2016).

FIGURE 3.84 Illustration of *Scoparia dulcis*.

Studies on the phytochemical and pharmacological profile of the SD plant have presented ammeline (Figure 3.85a) and scoparic acid D (Figure 3.85b), which along with other diterpenes, triterpenes, and flavonoids (Figure 3.4) also contribute to the plant's ability to control elevated BG levels (Sarkar et al. 2020, Mishra, Mishra et al. 2013, Paul, Vasudevan and Krishnaja 2017). Furthermore, in their review, Mishra et al. identified the bioactive compounds cirsimarin/cirsitakaoside (Figure 3.85c), cynaroside (Figure 3.85d), and stigmasterol (Figure 3.85e) as contributors to the plant's antidiabetic activity (Mishra, Behera et al. 2011). They further suggested that the compounds glutinol (Figure 3.85f) and coixol (Figure 3.85g), obtained from the aqueous decoctions, were found to be potent and active in insulin secretagogue activity (Pamunuwa, Karunaratne and Waisundara 2016, Sarkar et al. 2020).

The hypoglycaemic activity of the methanolic extract of the SD plant was investigated by Mishra et al., using both *in vitro* and *in vivo* models. Their *in vitro* experiments, which were aimed at studying the inhibitory effects of the methanolic extract on the two intestinal enzymes: α-amylase and α-glycosidase, confirmed its effectiveness in controlling PBG levels, while the results from their *in vivo* studies, using STZ-induced diabetic rats, revealed a significant arrest and decrease in the BG levels (at an extract dosage of 400 mg/kg.bw). This activity was found to be comparable to that of the control standard, glibenclamide (Mishra, Mishra et al. 2013).

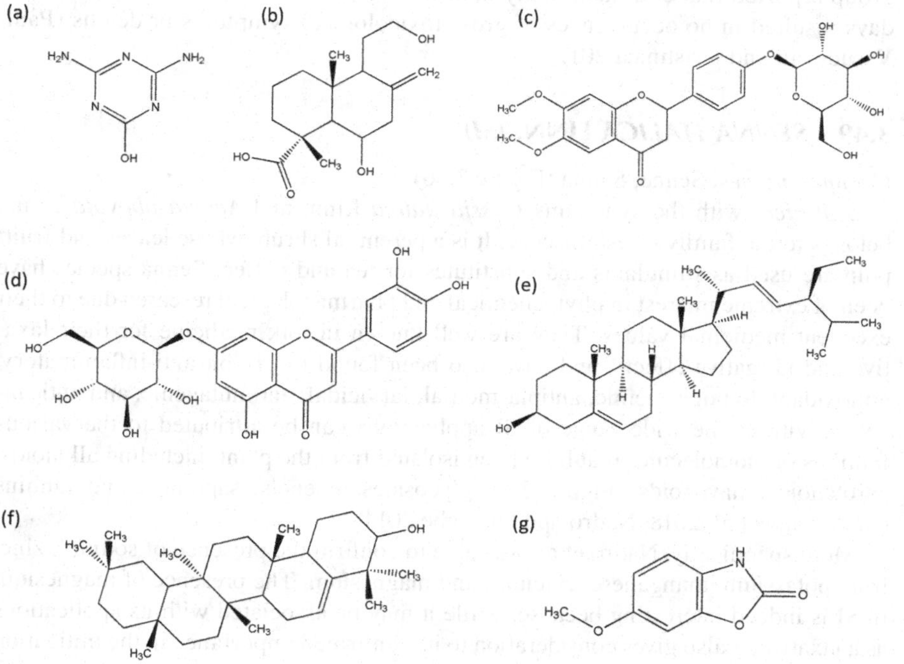

FIGURE 3.85 Structural formulae of isolated compounds from *Scoparia dulcis*: (a) ammeline, (b) scoparic acid D, (c) cirsimarin/cirsitakaoside, (d) cynaroside, (e) stigmasterol, (f) glutinol, and (g) coixol.

Pari and Latha investigated the effect of SD whole-plant extracts (aqueous, ethanolic, and chloroform) on key hepatic metabolic enzymes that have been identified in carbohydrate metabolism. The group reported that the aqueous extract showed superior antihyperglycaemic activity, in comparison to its organic extract competitors in the treated STZ-induced diabetic rodent subjects (Murti et al. 2012, Pari and Latha 2004). In a following study, Pari et al. demonstrated the ability of the SD plant extract to attenuate hyperglycaemia and also maintain optimal GSH levels in STZ-induced diabetic rats, by reducing the influx of glucose through the polyol pathway (Murti et al. 2012, Latha and Pari 2004). Adding to this, Latha et al. further presented the insulin secretagogue potential of the SD plant, when a 10 μg/mL dose was able to stimulate a 6.0 fold secretion of insulin from isolated mouse pancreatic tissue. Additionally, it was reported that the extract protected against STZ-mediated cytotoxicity (88%) and nitric oxide production, in the rat insulinoma cell lines (RINm5F) (Latha, Pari and Sitasawad et al. 2004).

Like all other potential botanical sources of therapy, concerns have been voiced regarding the plant's toxicity and safety in use. Two independent reviews have shown that SD extract therapy can be considered relatively safe (Sarkar et al. 2020, Paul, Vasudevan and Krishnaja 2017) as its aqueous extract did not produce any mortality up to the excessive oral dose level of 8 g/kg.bw in mice. Furthermore, the group reported that continued daily administration of the extract for a total of 30 days resulted in no occurrences of gross toxicological symptoms or deaths (Paul, Vasudevan and Krishnaja 2017).

3.49 *SENNA ITALICA* LINN. *(SI)*

Common names: Senne, Senna (Figure 3.86)

S. italica, with the synonyms *Cassia italica* Linn. and *Acacia obovata* Linn., belongs to the family Caeslpinaceae. It is a perennial shrub, whose leaves and fruit/pulp are used as stimulants and substitutes for tea and coffee. Senna species have been of extreme interest in phytochemical and pharmacological research due to their excellent medicinal values. They are well known in folk medicine for their laxative and purgative effects, and have also been found to exhibit anti-inflammatory, antioxidant, hypoglycaemic, antiplasmodial, larvicidal, antimutagenic, and anticancer activities. The wide range of SI applications can be attributed to the various families of biomolecules that have been isolated from the plant, including alkaloids, carotenoids, flavonoids (Figure 3.4), glycosides, phenols, saponins, and tannins (Malematja et al. 2018, Nadro and Onoagbe 2014).

More specifically, Nadro et al. was able to confirm the presence of sodium, zinc, iron, potassium, manganese, calcium, and magnesium. The presence of magnesium in SI is indeed interesting because, while it may be associated with its applications as a laxative, it also gives consideration to its confirmed importance in the mitigation of insulin resistance (Nadro and Onoagbe 2014).

A study by Malematja et al. evaluated the *in vitro* effects of SI's acetone leaf extracts on GLUT-4 translocation, expression, and adipogenesis in 3T3-L1 pre-adipocyte cells. The results showed a high anti-glycation effect attained at 10 mg/mL,

FIGURE 3.86 Illustration of *Senna italic.*

while at 25–200 μg/mL, there was no discerning increase in adipogenesis and lipolysis. Interestingly, the extract (at 100 μg/mL) was linked to the decrease in the expression levels of various adipokines, but had minimal effect on glucose uptake at 50–100 μg/mL, when used in combination with insulin. However, the combinatory and synergistic effect (using the aforementioned extract dose range along with insulin) was noted to influence both GLUT-4 translocation and its increase in expression. Of equal importance, toxicity studies revealed that the extract had no adverse effect on the cells' viability, which emphasized the plant's safety in use. In the end, the group was able to confirm the safe and potential use of SI as a viable agent for both the anti-glycation and the down-regulation of obesity-associated adipokines (Malematja et al. 2018).

3.50 *STACHYTARPHETA JAMAICENSIS* LINN. *(SJ)*

Common names: Blue porterweed, Blue snake weed, Bastard vervain, Brazilian tea, Jamaica vervain, Light-blue snakeweed (Figure 3.87)

S. jamaicensis is a member of the Verbenaceae family. It is a weedy, herbaceous plant, and has an average height of 60–120 cm. The plant is found mainly in the tropical regions of the Americas, as well as in the subtropical forests of Africa, Asia, Oceania, and the Caribbean (Liew and Yong 2016, Ezenwa, Igbe and Idu 2015, Estella, Obodoike and Esua 2020).

Estella et al. evaluated the antidiabetic activity in different organic fractions of the SJ leaves, using protocoled *in vitro* and *in vivo* experiments. They reported that the extract showed significant dose-dependent hypoglycaemic activity in both of their acute and chronic *in vivo* studies, yielding respective reductions in the BG levels by (25.40% and 55.80%) against glibenclamide (12.60% and 51.40%) in normal rats, and the BG levels by (22.00% and 86.90%) against glibenclamide (19.37% and 85.50%) in diabetic rats (Estella, Obodoike and Esua 2020).

FIGURE 3.87 Illustration of *Stachytarpheta jamaicensis.*

Another study by Ezenwa et al. showed that although the total reduction in BG levels using glibenclamide (at 10 mg/kg.bw) was higher than that of the plant extract at different time points of the experiment, the overall hypoglycaemic effect of the extract (at 400 mg/kg.bw) was comparable to that of the control standard (Ezenwa, Igbe and Idu 2015).

Similar therapeutic doses were described by Egharevba et al., who reported that daily oral administration of the methanolic and ethyl acetate leaf extracts (at 200 and 400 mg/kg.bw) significantly decreased the BG levels of STZ-induced diabetic rats, as compared to the untreated diabetic animals (Egharevba et al. 2019). Supportively, Rozianoor et al. demonstrated the ameliorative effect of the ethanolic leaf extract by showing its ability to attenuate the spiked BG levels in a population of male diabetic rats. The group proposed that the observed therapeutic effect was possibly due to the presence of the contained bioactive compounds: genipin (Figure 3.88a) and linolenic acid (Figure 3.88b) (Ma et al. 2013, Rozianoor, Eizzatie and Nurdiana 2014).

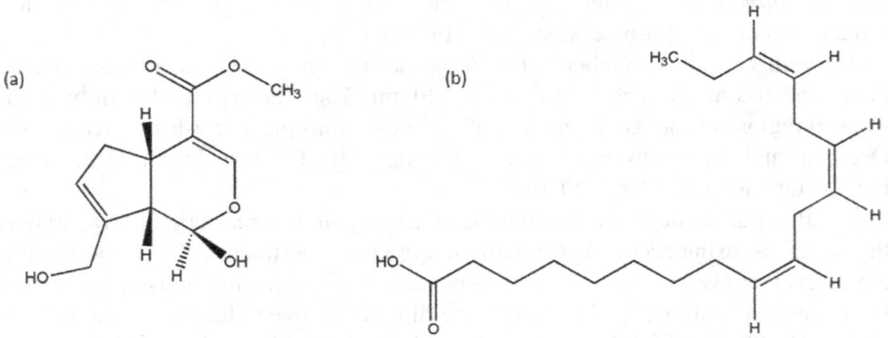

FIGURE 3.88 Structural formulae of isolated compounds from *Stachytarpheta jamaicensis*: (a) genipin and (b) linolenic acid.

Finally, results from various independent toxicity studies showed that even at high doses, the SJ leaf extracts influenced little or no observed abnormalities in the subjects' health and, as such, it was deemed safe for therapeutic use (Liew and Yong 2016, Ezenwa, Igbe and Idu 2015). It was noteworthy that in the acute toxicity and lethality tests that were performed by the Estella group, the LD_{50} was established to be >5000 mg/kg.bw, thereby underscoring the relative non-toxic nature of the SJ plant (Estella, Obodoike and Esua 2020).

3.51 *SYZYGIUM CUMINI* LINN. *(SC)*

Common names: Black plum, Goolab Jamun, Jambolan, Jamun, Java plum, Malabar plum (Figure 3.89)

S. cumini is a rapidly growing species that can reach heights of up to 30 m and can live more than 100 years. Its dense foliage provides shade and was initially grown for its ornamental value. At the base of the tree, the bark is rough and dark grey, becoming lighter grey and smoother higher up. The wood can be used to make furniture and other construction-related applications due to its water-resistant nature. Its leaves are pinkish when young and change to a leathery, glossy dark green with a yellow midrib, as they mature. The trees start flowering from March to April, and their flowers are fragrant and small (about 5 mm in diameter). The oblong fruits that develop resemble large berries and change from green to crimson red as they mature. Their taste has a combination of sweet and mildly sour flavours and they tend to discolour the tongue when eaten (Wikipedia 2021).

While in most other cases there is only anecdotal evidence for the antidiabetic properties of traditionally used medicinal plants, the SC plant has been extensively studied for its astringent, antidiarrhoeal, antidiabetic (Helmstädter 2008),

FIGURE 3.89 Illustration of *Syzygium cumini*.

antilipidemic (Jagetia 2018, Zulcafli et al. 2020), and antioxidant (Sharma et al. 2012) properties. Different parts of the jambolan tree, especially its fruits, seeds, and stem-bark exhibit promising activity against DM. These properties have been confirmed by several supporting experimental and clinical studies that we shall now further discuss. Tea prepared from SC leaves was reported to have antihyperglycaemic effects, while the stem-bark of the plant was found to induce the appearance of positive insulin-staining cells (in the epithelia of the pancreatic duct of treated animals) and influence a significant decrease in the BG levels, in treated mice (Ayyanar and Subash-Babu 2012).

Additionally, a study was carried out by Kumar et al. with the aim to isolate and characterize the putative antidiabetic compound from the SC seed. The group was successful in identifying the glucosamine, mycaminose (Figure 3.90a), which was then studied for its glucose-lowering efficacy in a population of STZ-induced diabetic rats. Here, the group reported that a dose of 50 mg/kg.bw (seed extract) was able to effect a significant reduction in the BG levels of the treated rats, which interestingly was similar to that of the control standard, glibenclamide (1.25 mg/kg.bw). Because of the known mechanism of action of the control standard, it was hypothesized that mycaminose followed a similar action pathway in stimulating insulin secretion from pancreatic β-cells (Kumar et al. 2008).

The ambiguity of the mechanism underlying SC's ameliorative action inspired Sharma and coworkers to investigate whether its aqueous seed extract (100, 200, or 400 mg/kg.bw) had any beneficial effects on insulin resistance, serum lipid profile, antioxidant status, and/or pancreatic β-cell damage in high-fat diet/STZ–induced diabetic rats. Sharma proudly demonstrated (through *in vivo* and cytology studies) that SC therapy (at a dose of 400 mg/kg.bw) was able to restore altered pathological

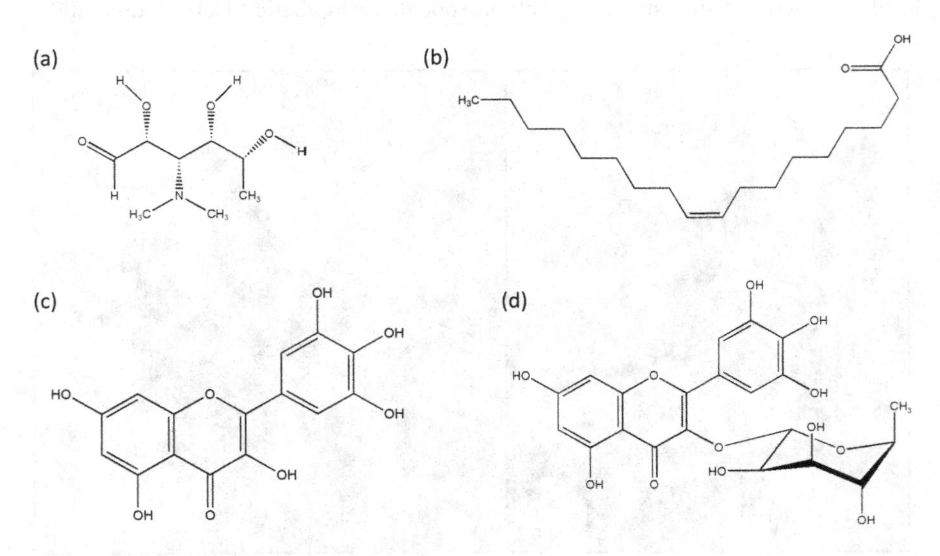

FIGURE 3.90 Structural formulae of isolated compounds from *Syzygium cumini*: (a) mycaminose, (b) oleic acid, (c) myricetin, and (d) myricitrin.

changes to pancreatic islets and β-cells and also influence an increase in PPARγ and PPARα protein expressions in hepatic tissue. In their discussion of the results, Sharma cited work by Anandharajan et al., who proposed that SC augments glucose uptake by upregulating the glucose transporter-4, PPARγ, and phosphatidylinositol 3-kinase in L6 myotubes. The increased levels of PPARγ were shown to influence both hypoglycaemic and hypolipidemic activity and also insulin-sensitizing actions in mild and severe diabetic rats. Another interesting observation by the group was that SC treatment in diabetic rats caused a significant decrease in serum tumour necrosis factor-alpha (TNF-α), which further supported its therapeutic role in T2DM. Sharma went on to justify their observations regarding the effects on PPARα activity, by attributing the attenuation of dyslipidaemia to the receptor's increased expression. In conclusion, they noted that single action or combinations of the phytochemicals (contained in SC seed extract), such as triterpenoids, anthocyanins, oleic acid (Figure 3.90b), essential oils, glycosides, saponins, and several members of the flavonoids (e.g. rutin [Figure 3.12e], quercetin [Figure 3.25g), myricetin (Figure 3.90c), and myricitrin (Figure 3.90d) were responsible for the plant's DM therapeutic action (Sharma et al. 2012).

In his review entitled "*Syzygium cumini* (L.) Skeels (Myrtaceae) against Diabetes – 125 Years of Research," Helmstaedter presented the controversial effects of SC therapy in clinical studies, as they were reported by independent research (Helmstädter 2008). Using the report from Sepaha and Bose (1956), he highlighted their observation that less than 50% of their test patients showed a positive response to SC therapy (Bose and Sepaha 1956). Unfortunately, these findings were further corroborated by studies carried out in India and Brazil and, as such, did not meet the modern criteria for clinical studies to be performed. However, Shrivastava and coworkers treated 28 "severely diabetic patients" with 4–24 g SC seed powder (in gelatine capsules) and reported a significant reduction in mean fasting (–18%) and postprandial (–32%) blood sugar levels. Unfortunately, several adverse drug-related reactions were reported in less than 20% of the patients, which included nausea, diarrhoea, and epigastric pain (Srivastava, Venkatarishna-Bhatt et al. 1983).

Similarly, another dissuading study by Teixeira et al. reported randomized patients (with T2DM) being treated with a tea that was prepared from SC leaves and its effect being gauged against that of the standard control drug, glyburide. It was found that FBG levels decreased significantly in participants who were treated with the allopathic drug, but showed no changes in those who were treated with either the SC tea or combinations with placebos. In the end, the researchers adamantly concluded that SC therapy was ineffective in treating complications related to T2DM (Teixeira et al. 2005).

Due to the opposing data being published on SC therapy, it is understandably confusing whether or not this form of therapy is effective. A closer look at Helmstaedater's reports revealed that the SC studies were dose responsive and time related, such that investigations that used the extracts at higher doses and for longer periods of time appeared to get more desirable responses than the low-dose, short-term experiments. This idea was well supported by Kohli and Singh, who reported a study on 30 patients with "uncomplicated" NIDDM. In their investigations, patients received 12 g SC seed powder in three divided doses for 3 months. An oral glucose tolerance test was performed every month and subjective parameters were monitored. There was a considerable and progressively increasing relief of symptoms such as polyuria, polyphagia, weakness, and weight loss. The results of the glucose tolerance test were significantly improved after 3 months of treatment. One- and two-hour values were reduced by up to 30% after 2 months of treatment compared to their control (chlorpropamide). Furthermore, they reported that the seed powder did not show any side effects to patient-health during their study (Kohli and Singh 1993)

3.52 *TAMARINDUS INDICA* LINN. *(TI)*

Common names: Tamarind, Tambran (Figure 3.91)

T. indica belongs to the Fabaceae family and is a long-lived fruit tree that bears fragrant flowers and fruit (Useful Tropical Plants n.d.). The fruit is easily recognized by its brittle exterior (fruit shell) and chewy, but sour pulp that is commonly used in the food industry (in the form of condiments, sauces, and as processed confectionaries). The tree and its edibles have been found to contain tannins, saponins, glycosides, flavonoids (Figure 3.4), and polyphenols (Bhadoriya et al. 2018) that are all well known for their associated medicinal properties, which range from anti-inflammatory, antioxidant, natural laxative, and antihyperglycaemic activity.

Bhadoriya et al. conducted a study on the antidiabetic potential of the TI seed coat in alloxan-induced diabetic rats. They used a hydro-ethanolic extract from the seed coat and administered doses of 100 and 250 mg/kg.bw to the diabetic rats daily for 14 days. After the treatment, their results indicated that both doses significantly

FIGURE 3.91 Illustration of *Tamarindus indica*.

reduced the BG levels in normoglycaemic and alloxan-induced hyperglycaemic rats, in relation to the control group (Bhadoriya et al. 2018). In glucose-loaded hyperglycaemic rats, both doses showed similar efficacy by significantly attenuating the BG levels as well. Complementary to these results was the observation that the body weight of the rodents had decreased during the administration of the extract.

While the results from the Bhadoriya research group seemed promising, results from two other groups were found to be quite contradicting (perhaps due to the seed coat being used in the aforementioned study and the seeds by themselves being used in the proceeding reports). Maiti et al. conducted a study on the antidiabetic effect of the aqueous extract of the TI seed on STZ-induced diabetic rats, over a period of 7 and 14 days. Unfortunately, the group reported that there were no observed changes in the FBG levels when daily doses of 80 mg/0.5 mL distilled water/100 g.bw were administered to the diabetic subjects (Maiti et al. 2004). Similarly in another study, Parvin et al. administered a seed powder mixture of 1.25 g/kg.bw/10 mL of water to STZ-induced diabetic rats for 24 hours. They too observed relatively paltry activity, whereby the serum glucose level was reduced by only 2.43% at 60 minutes and 7.68% at 120 minutes (Parvin et al. 2013).

It can be gathered that the potency of the seed extract when used as a form of diabetes therapy is inconclusive. However, further studies are encouraged to either support or refute the idea of its use as an alternative means of BG regulation. Additionally, due to the lack of available information, studies that explore the antidiabetic activity of the extracts from its leaves and bark should also be encouraged.

3.53 *TOURNEFORTIA HIRSUTISSIMA* LINN. *(TH)*

Common names: Chiggery grapes, Jigger Bush (Figure 3.92)

T. hirsutissima, of the Boraginaceae family, is a woody vine, with cylindrical pubescent stems, and bisexual flowers that are white or yellow to greenish and may produce five lobulated fruit. Its applications in ethnomedicine include the treatment of rheumatic, urinary, and also hyperglycaemic-related complications (Andrade-Cetto, Revilla-Monsalve and Wiedenfeld 2007).

FIGURE 3.92 Illustration of *Tournefortia hirsutissima.*

A study by Andrade-Cetto et al. compared the antidiabetic activities of the butanolic extracts (BE) and dehydrated aqueous extracts (WE) of TH, against those of glibenclamide. Using a population of 77 adult Wistar rats, the animals were equally divided into seven groups. Group 1 (non-diabetic control), received 1.5 mL of physiological NaCl solution, while Group 2 (diabetic control) also received the same volume of the physiological saline solution. The subjects of Group 3 were orally treated with glibenclamide (at a dose of 3 mg/kg.bw) and used as a standard control. For the remaining Groups 4–7, the test compounds were administered at increasing doses, in accordance with their method of extraction: Groups 4 and 5 received WE (20 and 80 mg/kg.bw) and Groups 6 and 7 received BE (8 and 80 mg/kg.bw), respectively. The results revealed that the effect of the administered extract was dose dependent, with the greatest efficacy observed using a dose of 80 mg/kg.bw. Appreciatively, it was found that at 80 mg/kg.bw, the extract activity was similar to the control-standard group (Andrade-Cetto, Revilla-Monsalve and Wiedenfeld 2007, Andrade-Cetto, Cabello-Hernández and Cárdenas-Vázquez 2015).

3.54 *URENA LOBATA* LINN. *(UL)*

Common names: Caesarweed, Congo jute, Cooze mahot, Kuze maho, Pink Chineseburr (Figure 3.93)

U. lobata is a member of the family: Malvaceae, and is traditionally used to treat several ailments including diabetes. It is a sub-shrub that has an average height of 0.6–3 m and a basal diameter of 0.7 m (Omonkhua and Onoagbe 2007), and is usually grown as annuals in tropical and temperate regions of Asia, Africa, and the Americas (Babu, Madhuri and Ali 2016).

Research showed that methanolic, ethanolic, and hot aqueous leaf extracts of UL exhibited DPP-4 inhibitory activity, thus supporting their effective use as diabetic therapeutic agents. Owing to these ameliorative effects, are the isolated compounds: β-sitosterol (Figure 3.10a), mangiferin (Figure 3.63a), and stigmasterol (Figure 3.85e)

FIGURE 3.93 Illustration of *Urena lobata*.

(Islam and Uddin 2017, Y. Purnomo, D. W. Soeatmadji et al. 2015, Y. Purnomo, D. W. Soeatmadji et al. 2015). It was further noted, by the Purnomo-led group, that the ethanolic extract showed stronger DPP-4 inhibitory effects than the aqueous extracts *in vitro,*) however, the opposite was observed when the studies were performed *in vivo*. The difference in activity was attributed to the greater solubility of the active compounds in the more polar medium, which also influenced better absorption within the gastrointestinal system (Y. Purnomo, D. W. Soeatmadji et al. 2015, Y. Purnomo, D. W. Soeatmadji et al. 2018).

In the same study by Purnomo et al., the oral administration of UL at doses of 250, 500, and 1000 mg/kg.bw to diabetic rats prevented degradation of GLP-1 by inhibiting DPP-4 activity (Y. Purnomo, D. W. Soeatmadji et al. 2015). Later, the group reported an increase in GLP-1 bioavailability through UL administration, and suggested that the therapy could lead to an improvement in the structure and function of the islet β-cells (Y. Purnomo et al. 2017). Other groups have successfully demonstrated the comparable therapeutic effects of the UL root extracts versus leaf extracts in diabetic animal models, and have shown that the extracts from the roots also exhibited potent antidiabetic/hypoglycaemic effects in their test subjects (Babu, Madhuri and Ali 2016, Islam and Uddin 2017, Omonkhua and Onoagbe 2017, Y. Purnomo, D. W. Soeatmadji et al. 2015, Y. Purnomo, D. W. Soeatmadji et al. 2018, Wahyuningsih and Purnomo 2018).

Finally, Omonkhua and Onoagbe investigated the occurrence of any possible toxic events that could be associated with UL therapy in healthy rabbits. They found that although there was an initial ill response on the rabbits' hepatocytes as well as preliminary bile obstruction, the effects were not sustained (Omonkhua and Onoagbe 2011). Furthermore, the UL extracts did not appear to exert any form of oxidative damage to normal (healthy) rabbits, with respect to the liver and pancreatic MDA levels; as a matter of fact, they even seemed to be protective against lipid peroxidation (Omonkhua and Onoagbe 2007, Omonkhua and Onoagbe 2012).

3.55 *ZINGIBER OFFICINALE* LINN. *(ZO)*

Common names: Halia, Common Ginger, Canton Ginger, Stem Ginger, Ginger (Figure 3.94)

Z. officinale belongs to the Zingiberaceae family, along with its relative *C. longa* (turmeric). It is popularly known as ginger in the Caribbean and is a perennial plant that grows below or on the surface of the soil. In the Caribbean, ZO is consumed as a drink (tea or soft drink), and used as a seasoning for meats and foods or as a spice for desserts. More importantly, ZO is popularly known to have strong medicinal properties, used to reduce arthritis, migraines, abdominal pain, and nausea (I. A. Ross 2005). Furthermore, *in vitro* studies on the isolated active compounds gingerol (Figure 3.95a) and shogaol (Figure 3.95b) have confirmed their antidiabetic activity, thus making the rhizome a good option for the maintenance of normoglycaemic levels (Rani et al. 2011).

In our searches, we have found several animal and human studies that aimed to confirm the hypoglycaemic activity of the plant. In a clinical study, Mahabir and

FIGURE 3.94 Illustration of *Zingiber officinale*.

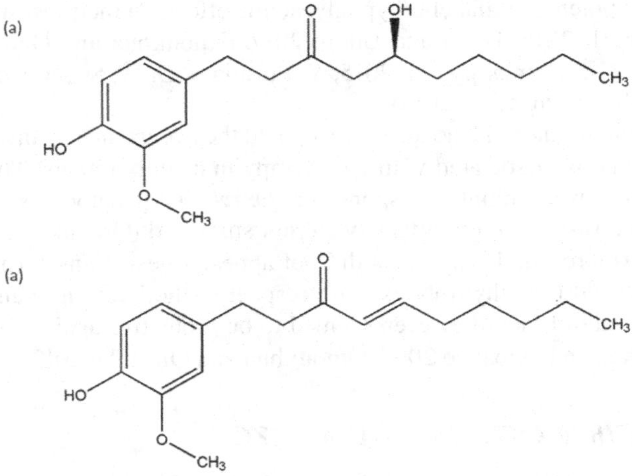

FIGURE 3.95 Structural formulae of isolated compounds from *Zingiber officinale*: (a) gingerol and (b) shogaol.

Gulliford, for example, used a group of 264 diabetic individuals to evaluate the diabetic activity of ginger and other herbal medicines. They concluded that ginger, among other herbs used, was able to restore euglycaemic levels in individuals who had diabetes, and were consuming the hot water extract (ginger tea) as a form of therapy (Mahabir and Gulliford 1997).

Alternatively, in 2006, Al-Amin et al. studied the hypoglycaemic potentials of ZO in STZ-induced diabetic rats. A 500 mg/kg.bw dose of an aqueous extract of raw ginger was administered to these rats daily for 7 weeks. After the 7-week treatment period, the results showed that the therapeutic dose produced a significant reduction in both the BG levels, as well as the urine protein levels in the diabetic rats (Al-Amin et al. 2006).

In this final report, we saw the ethanolic extract of the rhizome being administered to diabetic rabbits, at a dose of 100 mg/kg.bw over 3 weeks. The research group reported that the therapy resulted in a significant reduction in the BG levels, when compared to the untreated animals, and thus confirmed the potential use of the ZO extract as a viable and relatively safe means to attenuate hyperglycaemia in DM subjects (Mascolo et al. 1989).

TABLE 3.1
Summary of the Medicinal Plants with Their Respective Dosages and Mechanisms of Action

Common Name	Scientific Name	Extract Type and Therapeutic Dose	Mechanism of Antidiabetic Action (in Animal Models)	Observed Degree of Antidiabetic Activity
Aam, Manako, Mango	*Mangifera indica*	Leaf and stem bark extract: (250 mg/kg.bw)	1) Reduction in glucose absorption, in the GI tract. 2) Stimulation of β-cell release of insulin.	Good efficacy when compared to glibenclamide.
Aanabahe-hindi, Buah papaya, Papaw, Papaya, Pawpaw	*Carica papaya*	Leaf extract: (400 and 200–600 mg/kg.bw)	1) Enhanced secretion of insulin from the pancreatic β-cells. 2) Alpha-amylase inhibitory activity.	Dose dependent with very good correlation with gibenclamide, at the highest dosages.
Ach, Achi, Awltree, Bo-aal, Cheese Fruit, Hog apple, Indian mulberry, Noni, Nono, Nonu	*Morinda citrifolia*	Fruit juice (2 mL/kg.bw)	Promotes insulin sensitivity in adipose tissue and muscles.	Good efficacy when compared to glibenclamide.
Achiote, Annatto, Ookoo plant, Roucou	*Bixa orellana*	Seed extract: (80 and 540 mg/kg.bw)	Inconclusive mechanism.	Moderate.
Ajmod, Celeriac, Celery, Leaf celery, Root celery, Wild celery	*Apium graveolens*	Seed extract: (425 mg/kg.bw) Leaf extract: (150 mg/kg.bw)	Direct stimulation of glycolysis in peripheral tissues, and a reduction in the absorption of glucose from the gastrointestinal tract.	Satisfactory reduction in BG levels relative to untreated rodent groups.
Alaf-e-Mar, Alcaparro, Caper bush, Cappero, Kebbar, Wild Watermelon	*Capparis spinosa*	Fruit extract: (20 mg/kg.bw)	1) Reduction of carbohydrate absorption in the small intestine; inhibition of gluconeogenesis in the liver. 2) Enhancement of the uptake of glucose by tissues and β-cell conservation.	Good correlation with sodium vanadate.

(*Continued*)

TABLE 3.1 (CONTINUED)

Summary of the Medicinal Plants with Their Respective Dosages and Mechanisms of Action

Common Name	Scientific Name	Extract Type and Therapeutic Dose	Mechanism of Antidiabetic Action (in Animal Models)	Observed Degree of Antidiabetic Activity
Aloe, Aloe vera, Alovis, Sabila	*Aloe barbadensis*	Leaf pulp extract: (500 mg/kg) Leaf gel extract: (10 mL/kg)	Control of carbohydrate metabolizing enzyme activity.	Good efficacy when compared to glibenclamide.
Asian Ginseng, Chinese Ginseng, Guigai, Jiln Ginseng, Korean Ginseng, Ninjin, Oriental Ginseng, Panax schinseng, Red Ginseng, American ginseng, Canadian ginseng, Ginseng	*Panax ginseng* *Panax quinquefolius*	Root extract: (100–600 mg)	1) Increased PPAR-γ expression and AMP-activated protein kinase phosphorylation in liver and muscle. 2) Decrease in carbohydrate absorption rate into the portal hepatic circulatory system. 3) Increased glucose transport and uptake. 4) Increased glycogen storage. 5) Modulation of insulin secretion.	Satisfactory reduction in BG levels, relative to untreated patients.
Asthma plant, Asthma weed, Soro, Garden spurge, Snakeweed	*Euphorbia hirta*	Leaf and stem extract: (200–400 mg/kg.bw)	1) Stimulation of insulin effects from remnant β-cells. 2) Inhibition of α-amylase activity.	Good efficacy when compared to glibenclamide.
Billygoat-weed, Tropical Whiteweed, Zeb-a-fam	*Ageratum conyzoides*	Shoot extract: (100–400 mg/kg.bw) Leaf extract: (100–500 mg/kg.bw) Fractionated leaf extract: (88 mg/kg.bw)	1) Inhibition of the GLUT-2 transporter (in the intestinal mucosa), and the α-glucosidase enzyme (in the small intestine). 2) Proposed reduction in glucose absorption and prevention of carbohydrate breakdown.	Good efficacy when compared to glibenclamide.

(Continued)

TABLE 3.1 (CONTINUED)
Summary of the Medicinal Plants with Their Respective Dosages and Mechanisms of Action

Common Name	Scientific Name	Extract Type and Therapeutic Dose	Mechanism of Antidiabetic Action (in Animal Models)	Observed Degree of Antidiabetic Activity
Bindi, Lady Fingers, Ochro, Okra, Okro	*Abelmoschus esculentus*	Powdered peel: (100 mg/kg.bw) Powdered seed: (200 mg/kg.bw) Fruit extract: (100–400 mg/kg.bw)	Exhibits α-glucosidase-inhibiting activity.	Good efficacy when compared to metformin.
Bitter Melon, Bitter Gourd, Bitter Squash, Balsam-pear, Karela, Carailla, Carille Cerasee, Coraillie	*Momordica charantia*	Wet fruit extract: (500 mg/kg.bw) Dehydrated fruit extract: (20 mg/kg.bw)	1) Stimulation of insulin release. 2) Stimulation of peripheral and skeletal muscle glucose utilization. 3) Inhibition of intestinal glucose uptake. 4) Inhibition of adipocyte differentiation. 5) Suppression of key gluconeogenic enzymes. 6) Stimulation of key enzyme of HMP pathway. 7) Preservation of islet β-cells and their functions. 8) Regulation of AMP-activated protein kinase α and glucagon-like peptide-1 via bitter taste receptor signalling.	Good efficacy when compared to metformin, and tolbutamide. Good efficacy when compared to glibenclamide.

(Continued)

TABLE 3.1 (CONTINUED)

Summary of the Medicinal Plants with Their Respective Dosages and Mechanisms of Action

Common Name	Scientific Name	Extract Type and Therapeutic Dose	Mechanism of Antidiabetic Action (in Animal Models)	Observed Degree of Antidiabetic Activity
Black Jack, Cobbler's pegs, Farmers friend, Hairy beggarticks, Needle bush, Spanish needle, Sticky beaks	*Bidens pilosa*	Whole plant extract: (50–400 mg/kg.bw)	1) Stimulation of insulin secretion from pancreatic β-cells. 2) Additional benefits of restoration or protection of the pancreatic cells.	Good efficacy when compared to glibenclamide.
Black plum, Goolab Jamun, Jambolan, Jamun, Java plum, Malabar plum	*Syzygium cumini*	Seed extract: (50–400 mg/kg.bw)	1) Stimulation of insulin secretion from pancreatic β-cells. 2) Influences an increase in PPARγ and PPARα protein expressions in hepatic tissue.	Good efficacy when compared to glibenclamide and chlorpropamide.
Blue porterweed, Blue snake weed, Bastard vervain, Brazilian tea, Jamaica vervain, Light-blue snakeweed	*Stachytarpheta jamaicensis*	Leaf extract: (200–400 mg/kg.bw)	Activates IRS-1, PI3-K, and downstream signalling pathway and increases concentrations of calcium, resulting in GLUT4 translocation and glucose uptake increase.	Good efficacy when compared to glibenclamide.
Bright eyes, Cape periwinkle, Graveyard plant, Madagascar periwinkle, Old maid, Pink periwinkle, Rose periwinkle	*Catharanthus roseus*	Leaf extract: (100–1000 mg/kg.bw)	Enhancement of insulin secretion from the β-cells of Langerhans.	Dose dependent with 500 mg/kg.bw being most efficient.

(Continued)

TABLE 3.1 (CONTINUED)
Summary of the Medicinal Plants with Their Respective Dosages and Mechanisms of Action

Common Name	Scientific Name	Extract Type and Therapeutic Dose	Mechanism of Antidiabetic Action (in Animal Models)	Observed Degree of Antidiabetic Activity
Brown mustard, Mustard greens, Chinese mustard, Leaf mustard, Rai	*Brassica juncea*	Seed extract: (250–450 mg/kg.bw)	1) Increases the activity of glycogen synthetase. 2) Stimulation of insulin secretion from pancreatic β-cells.	Dose dependent with very good results at the higher dosages.
Bulb Onion, Common Onion	*Allium cepa*	Bulb extract: (25–200 g)	Increasing the concentration of free insulin in the body.	Good efficacy when compared to tolbutamide.
Caesarweed, Congo jute, Cooze mahot, Kuze maho, Pink Chineseburr	*Urena lobata*	Leaf extract: (250–1000 mg/kg.bw)	Exhibits DPP-4 inhibitory activity.	Good efficacy when compared to the control.
Calabacero, Calabash, Kalbas, Miracle fruit, Rum tree, Totumo	*Crescentia cujete*	Fruit extract: (10,000 ppm = 10 mg/L)	Stimulation of insulin secretion from pancreatic β-cells.	Good efficacy when compared to glibenclamide.
Caripeelay, Carypoulay, Curry leaf, Curry patta, Kaddi patta	*Murraya koenigii*	Leaf extract: (200–400 mg/kg.bw)	Preservation of pancreatic β-cells.	Good efficacy when compared to glibenclamide.
Chhui-mui, Humble plant, Lajja, Laajvanti, Teemarie, Ti-Marie, Touch-me-not plant, Sensitive plant, Shameful plant	*Mimosa pudica*	Leaf extract: (300–600 mg/kg.bw) Seed extract: (100–400 mg/kg.bw)	Exhibits inhibitory action against α-amylase and α-glucosidase enzymes.	Good efficacy when compared to metformin. Good efficacy when compared to glibenclamide.

(*Continued*)

TABLE 3.1 (CONTINUED)
Summary of the Medicinal Plants with Their Respective Dosages and Mechanisms of Action

Common Name	Scientific Name	Extract Type and Therapeutic Dose	Mechanism of Antidiabetic Action (in Animal Models)	Observed Degree of Antidiabetic Activity
Chiggery grapes, Jigger Bush	*Tournefortia hirsutissima*	Stem extract: (8–80 mg/kg.bw)	Inconclusive mechanism.	Dose dependent with very good correlation with gibenclamide at the highest dosages.
Cilantro, Coriander, Coriandre, Dhania	*Coriandrum sativum*	Seed extract: (200 mg/kg.bw) Leaf extract: (200–800 mg/kg.bw)	1) Influences enhanced glucose transport, glucose oxidation, and glycogenesis. 2) Exhibits inhibitory action against α-glucosidase enzymes.	Good efficacy when compared to glibenclamide.
Citron grass, Citronela, Citronella grass, Fever grass, Lemon grass	*Cymbopogon citratus*	Leaf extract: (125–500 mg/kg.bw) Essential oil: (800 mg/kg.bw)	1) Exhibits inhibitory action against α-glucosidase and α-amylase enzymes. 2) Observed improvement in insulin resistance.	
Clove basil, African basil, Wild basil	*Ocimum gratissimum*	Leaf extract: (200–400 mg/kg.bw)	Stimulation of insulin secretion from pancreatic β-cells.	Good efficacy when compared to the group that was treated with insulin.
East Indian basil, Holy basil, Krishna Toolsie, Tulasi, Tulsi	*Ocimum tenuiflorum*	Leaf extract: (250–500 mg/kg.bw)	Stimulation of insulin secretion from pancreatic β-cells.	Good efficacy when compared to glibenclamide.
Common Lantana, Lantana Red Sage Shrub, Verbena Yellow Sage, Caraquite, Cariaquito, West Indian lantana	*Lantana camara*	Fruit extract: (100–200 mg/kg.bw) Leaf extract: (200–500 mg/kg.bw)	Exhibits inhibitory action against α-amylase enzymes.	Good efficacy when compared to glibenclamide.

(*Continued*)

TABLE 3.1 (CONTINUED)
Summary of the Medicinal Plants with Their Respective Dosages and Mechanisms of Action

Common Name	Scientific Name	Extract Type and Therapeutic Dose	Mechanism of Antidiabetic Action (in Animal Models)	Observed Degree of Antidiabetic Activity
Cucumber	*Cucumis sativus*	Fruit juice: (400 g supplement) Fruit extract: (400 mg/kg.bw)	Inconclusive mechanism.	Good efficacy with the juice alone. Good efficacy when compared to glibenclamide.
Curcuma, Haldi, Hardi, West Indian saffron, Safflower, Turmeric, Yellow Ginger	*Curcuma longa*	Root extract: (80–300 mg/kg.bw)	1) Influences an increase in PPARγ activity. 2) Inhibition of inflammatory cytokines, such as MCP and TNF-α along with the induction of AMPK through the inhibition of MAPK.	Satisfactory reduction in BG levels relative to untreated rodent groups.
Dogoyaro, Indian lilac, Margosa tree, Neem, Nimtree	*Azadirachta indica*	Leaf extract: (250–400 mg/kg.bw)	1) Influences increase in glucose-6-phosphate dehydrogenase activity as well as hepatic and skeletal muscle glycogen content. 2) Proposed to influence the regeneration of damaged pancreatic β-cells.	Satisfactory reduction in BG levels relative to untreated rodent groups.

(*Continued*)

TABLE 3.1 (CONTINUED)

Summary of the Medicinal Plants with Their Respective Dosages and Mechanisms of Action

Common Name	Scientific Name	Extract Type and Therapeutic Dose	Mechanism of Antidiabetic Action (in Animal Models)	Observed Degree of Antidiabetic Activity
Drumstick tree, Horseradish tree, Moringa, Mother's best friend, Saijan, West Indian ben	*Moringa oleifera*	Leaf extract: (66 mg/kg.bw)	1) Decreased gluconeogenesis in the liver. 2) Increased insulin sensitivity and secretion. 3) Increased insulin sensitivity and secretion. 4) Inhibition of glucose uptake in the liver. 5) Increased glucose uptake in the liver and muscles.	Good efficacy when compared to metformin.
Fig-leaf gourd, Malabar gourd, Black seed squash, Cidra	*Cucurbita ficifolia*	Fruit extract: (300 mg/kg.bw)	Proposed to influence the regeneration of damaged pancreatic β-cells and increase plasma insulin levels.	Good efficacy when compared to tolbutamide.
Pumpkin, Field pumpkin, Ozark melon, Texas gourd	*Cucurbita pepo*	Seed powder (mixed with flax seed powder): (2–3 g/kg.bw)		Good efficacy when compared to tolbutamide.
Gale of the wind, Stonebreaker, Seed-under-leaf	*Phyllanthus amarus*	Leaf and stem extract: (200–1000 mg/kg.bw)	Influence a decrease in glucose-6-phosphatase and fructose-1-6-disphosphatase action and a simultaneous increase in the activity of glucokinase.	Good efficacy when compared to the control.
Garlic, Lahasun	*Allium sativum*	Bulb extract: (1–10 mL/kg.bw)	Improve plasma lipid metabolism and plasma antioxidant activity.	Similar efficacy when compared to diazepam and glibenclamide.

(*Continued*)

TABLE 3.1 (CONTINUED)
Summary of the Medicinal Plants with Their Respective Dosages and Mechanisms of Action

Common Name	Scientific Name	Extract Type and Therapeutic Dose	Mechanism of Antidiabetic Action (in Animal Models)	Observed Degree of Antidiabetic Activity
Good Luck, Resurrection plant, Wonder of the World	*Bryophyllum pinnate*	Leaf extract: (200 and 25–800 mg/kg.bw)	Exhibits α-glucosidase inhibiting activity.	Good efficacy when compared to glibenclamide and chlorpropamide respectively.
Gotu Kola, Asiatic Pennywort, Spade leaf	*Centella asiatica*	Whole plant extract: (1200 mg/kg.bw)	Exhibits α-glucosidase inhibiting activity.	Good correlation with metformin.
Grapefruit, Paradise Citrus	*Citrus paradisi*	Pure juice: (3.0 mL/kg.bw)	Influences the inhibition of protein and lipid catabolism, which is associated with insulin deficiency and increases hepatic glycogen content.	Good efficacy in T2DM models but not in T1DM.
Halia, Common Ginger, Canton Ginger, Stem Ginger, Ginger	*Zingiber officinale*	Root extract: (100–500 mg/kg.bw)	Associated with increased production of insulin.	Good efficacy when compared to tolbutamide.
Herbe á pic, Jackass Bitters, Zeb-a-pique	*Neurolaena lobata*	Whole plant extract: (5000 mg/kg.bw)	Inconclusive mechanism.	Good efficacy when compared to the control.
Hibiscus Rose mallow, Shoe black plant, Shoe flower plant, Tulipan	*Hibiscus rosa-sinesis*	Flower extract: (200–500 mg/kg.bw)	Associated with increased production of insulin.	Good efficacy when compared to glibenclamide.
Klip dagga, Christmas candlestick, Lion's ear, Chandilay bush	*Leonotis nepetifolia*	Leaf extract: (250–500 mg/kg.bw)	Associated with increased production of insulin.	Good efficacy when compared to glibenclamide.

(*Continued*)

TABLE 3.1 (CONTINUED)
Summary of the Medicinal Plants with Their Respective Dosages and Mechanisms of Action

Common Name	Scientific Name	Extract Type and Therapeutic Dose	Mechanism of Antidiabetic Action (in Animal Models)	Observed Degree of Antidiabetic Activity
Lemon	*Citrus limon*	Pure juice: (0.2–0.6 mL/kg.bw) Peel extract: (250–500 mg/kg.bw)	Amelioration of diabetes-associated conditions in insulin-resistant models.	Good efficacy when compared to glibenclamide.
Licorice weed, Goatweed, Scoparia-weed, Sweet-broom	*Scoparia dulcis*	Leaf and stem extract: (200–400 mg/kg.bw)	1) Exhibits inhibitory action against α-glucosidase and α-amylase enzymes. 2) Influences insulin secretagogue activity.	Good efficacy when compared to glibenclamide.
Lime, Key Lime, Mexican Lime, Mexican Thornless Key Lime	*Citrus aurantiifolia*	Pure juice: (50% dilution, 5 mL/kg.bw) Leaf essential oil: (100 mg/kg.bw)	1) Inhibition in the activities of gluconeogenic enzymes: glucose 6-phosphatase and fructose-1,6-bisphosphatase. 2) Increase in the activity of glucokinase.	Excellent correlation with metformin.
Orange, Sweet orange	*Citrus sinensis*	Peel extract: (12.5–100 mg/kg.bw) Bark extract: (400 mg/kg.bw)	Amelioration of diabetes-associated conditions in insulin-resistant models.	Good efficacy when compared to glibenclamide.
Pepper elder, Silverbush, Rat-ear, Shining bush plant, Man to Man	*Peperomia pellucida*	Dried leaf powder: (10%–20% w:w food supplement)	Enhanced secretion of insulin from the pancreatic β-cells.	Good efficacy when compared to glibenclamide.
Saint John's bush	*Justicia secunda*	Leaf extract: (2.5–3 g/kg.bw)	Exhibits inhibitory action against α-glucosidase enzymes.	Good efficacy when compared to glibenclamide.

(*Continued*)

TABLE 3.1 (CONTINUED)
Summary of the Medicinal Plants with Their Respective Dosages and Mechanisms of Action

Common Name	Scientific Name	Extract Type and Therapeutic Dose	Mechanism of Antidiabetic Action (in Animal Models)	Observed Degree of Antidiabetic Activity
Santa-Maria, Santa Maria feverfew, Whitetop weed, Famine weed, Whitehead broom	*Parthenium hysterophorus*	Flower extract: (100 mg/kg.bw)	Inconclusive mechanism.	Good efficacy when compared to glibenclamide.
Senne, Senna	*Senna italica*	Leaf extract (only found in *in vitro* studies: 10–200 µg/mL)	Proposed to influence both GLUT-4 translocation and its increase in expression.	
Siam weed, Christmas bush, Jack in the Box, Devil Weed, Communist Pacha, Common Floss Flower, Rompe Saragüey	*Chromolaena odorata*	Leaf extract: (200–400 mg/kg.bw) Root extract: (300–600 mg/kg.bw)	1) Reduction of carbohydrate absorption in the small intestine; inhibition of gluconeogenesis in the liver. 2) Enhancement of the uptake of glucose by tissues and β-cell conservation.	Dose dependent with very good correlation with gibenclamide at the highest dosages.
Soursop, Prickly Custard Apple, Graviola, Guanabana	*Annona muricata*	Seed extract: (600 and 800 mg/kg.bw)	1) Protection and preservation of pancreatic β-cell integrity. 2) Exhibits inhibitory action against FOXO1 protein.	Satisfactory reduction in BG levels relative to untreated rodent groups.
Tamarind, Tambran	*Tamarindus indica*	Seed coat extract: (100 and 250 mg/kg.bw)	Exhibits α-glucosidase inhibiting activity.	Good efficacy, when compared to glipizide.

(Continued)

TABLE 3.1 (CONTINUED)
Summary of the Medicinal Plants with Their Respective Dosages and Mechanisms of Action

Common Name	Scientific Name	Extract Type and Therapeutic Dose	Mechanism of Antidiabetic Action (in Animal Models)	Observed Degree of Antidiabetic Activity
Trumpet tree, Pop-a-gun, Tree-of-laziness, Snakewood tree	*Cecropia obtusifolia*	Leaf extract: (150 mg/kg.bw)	Exhibits α-glucosidase and glucose-6-phosphatase inhibiting activity.	Dose dependent with very good correlation with glibenclamide at the highest dosages.
Bacano, Bois Cano, Trumpet Tree, Snakewood, Congo pump, Wild pawpaw, Pop-a-gun	*Cecropia peltata*	Leaf extract: (200 mg/kg.bw)		
Water mint	*Mentha aquatica*	Whole plant extract: (100 mg/kg.bw)	Inconclusive mechanism; however, it is thought to exhibit α-glucosidase inhibiting activity as one of its mechanisms of action.	Good efficacy when compared to glibenclamide.
Wild mint	*Mentha arvenisis*	Whole plant extract: (50 mg/kg.bw)		Good efficacy when compared to acarbose.
Pudina, Spearmint	*Mentha spicata*	Whole plant extract: (300 mg/kg.bw)		Good efficacy when compared to glibenclamide.
Peppermint	*Mentha peperita*	Whole plant extract: (290 mg/kg.bw)		Good efficacy when compared to the control.

4 Metallopharmaceuticals

V
R
CrPic
You Know...You're wasting your time with that Research.
Metals will Never be Accepted as Medicine!

The failures to understand and follow the progress of the research behind many coordination complexes have led to very harsh and unfounded criticisms against chemists, especially by those in the medical field. Yet, in what could be considered hypocrisy, the coordination complex, cisplatin, is still recognized as one of the most successful and widely used medicines of this day.

4.1 INTRODUCTION

4.1.1 Overview and Significance

Metallopharmaceuticals or "metallo-drugs" fall under the umbrella of "alternative medicines" and show promise in the prevention, diagnosis, and treatment of an extensive range of ailments. Metal complexes exhibit pharmacological effects by reacting with biological molecules in a cell, with proteins and nucleic acids being the main targets for these agents. The interest in these relatively novel therapeutants has been justified by their demonstrated superiority in action over non-chelated compounds, which promise to serve the same purpose.

DOI: 10.1201/9781003322429-4

4.1.2 The Challenge

Nature has developed efficient defense mechanisms for mitigating the effects of metallo-drugs, by either sequestering and eliminating the toxic metal ion or repairing the damage done to critical targets such as DNA. Noting that the human body has tens of thousands of different proteins, involved in a variety of different catalytic, transport, and structural roles, if a metallo-drug complex were to attach itself to a part of an enzyme that is critical for catalytic function, the activity of the enzyme could be impaired. Further to this, other proteins have metal ions in parts of their structure that are important for maintaining protein structure and conformation. Again, if one or more of these metal ions is replaced with a metal ion supplied by a drug molecule having a coordination geometry that is different from the naturally occurring metal ion, the protein could lose its function due to this allosteric change (Dabrowiak 2017).

From the standpoint of coordination chemistry, only 7 of the 20 common amino acids found in proteins have donor atoms in their side chains that present them as potential targets for metallo-drugs. Unlike purely organic-based drugs, metal complexes that are used for treating and diagnosing diseases are exposed to different physiological environments in the body that could cause changes in their chemical composition as they travel from the site of administration to target molecules in a cell (Dabrowiak 2017).

The road from the laboratory bench to the clinic for all new drugs is marked with many pitfalls. Only a tiny fraction of new compounds makes it over the regulatory hurdles and is approved by the respective agencies such as the European Agency for the Evaluation of Medicinal Products (in Europe), the Food and Drug Administation (FDA; in the United States), and the Ministry of Health, Labor, and Welfare (in Japan). The approval process is a justifiably thorough one and, as a result, is known to be quite lengthy. Unfortunately, this has been a major deterrent to many scientists, due to fear of failure and wasted time.

Understandably, researchers have looked to the library of natural products and chosen suitable ligands that, by themselves, have exhibited desirable activities against particular diseases, and strategically mixed them with metals of interest. Although there is no guarantee that the result will be favourable, there have been instances where the resulting metal complex displayed higher levels of efficacy and lower levels of toxicity due to the nature of the ligand (Dabrowiak 2017, Crans et al. 2019, Adachi, Yoshida et al. 2004). Furthermore, many more metal-centred compounds have been synthesized by redesigning the existing chemical structure through ligand modification or building an entirely new compound with an enhanced safety and cytotoxic profile (Thompson and Orvig 2006, Rambaran et al. 2020). However, because of the increased emphasis on the clinical relevance of metal-based complexes, only a few of these drugs have made it to clinical trials and many more are awaiting ethical approval to join them.

With regard to the development of antidiabetic compounds, we have found that much work has been done using natural product isolates, such as allixin (Figure 4.1a), dipicolinic acid (Figure 4.1b), and maltol (Figure 4.1c), due to their known beneficial activities in various metabolic pathways (Dabrowiak 2017).

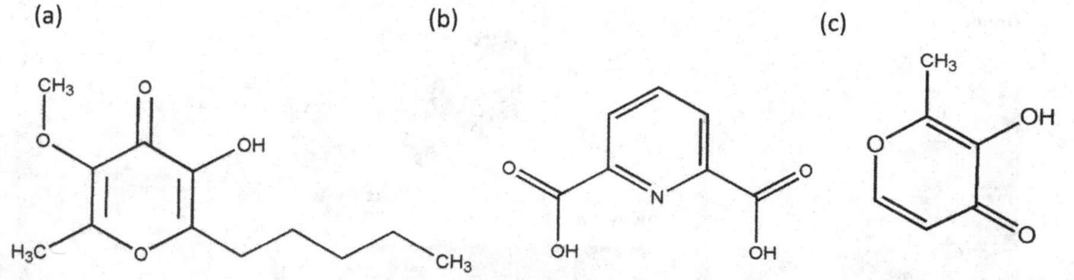

(a) (b) (c)

FIGURE 4.1 Structural formulae of natural product isolates: (a) allixin, (b) dipicolinic acid, and (c) maltol.

Fondly dubbed by their developers as "insulin mimetics" or "insulin enhancers," these metallo-drugs are known to effect the hypoglycaemic action in diabetic conditions through multiple intracellular pathways. Among them, vanadium has garnered much attention due to its ability to attenuate elevated blood sugar levels in both human and animal type-1 diabetes mellitus (T1DM) and type-2 diabetes mellitus (T2DM) subjects. Yet, like natural products, these drugs have also struggled for acceptance among the medical fraternity, due to incomplete research behind them. Pioneering work done by the Sakurai-led group and the Crans/Willsky and Orvig/McNeill teams continue to encourage and empower researchers to not give up, despite the many detracting opinions that may be directed at them. The McNeill/Orvig group inspired many with their development of the lead compound: bis(ethylmaltolato) oxovanadium(IV) (BEOV), which was successfully taken through Phases 1 and 2 clinical trials. Unfortunate rumours of the drug failing in the trials followed its shocking discontinuation into Phase 3 studies. However, it was revealed that a soon-to-be expired patent was the true cause for its withdrawal, due to there being no potential for generating revenue on these otherwise promising antidiabetic therapeutants.

4.1.3 The Future for Insulin-Enhancing Drugs?

Because there are large differences between having a compound being board approved as a drug and being licensed as a nutritional additive, metallopharmaceuticals have been allowed into the market as supplements, where the requirements for such are much less strict. Currently, several metals such as vanadium, chromium, and zinc exist as nutritional additives in either their salt or complex forms, and despite their mechanisms remaining obscure, are marketed as proponents for diabetes management and weight loss (Staff 2018). In instances like these, where available information on their respective formulations is limited, researchers must examine the publications reporting human and animal studies, as well as draw from the many commercial forms of the compound (Crans et al. 2019).

4.2 VANADIUM

The element vanadium is classified as a first-row transition metal, with the symbol V and atomic number 23 (Figure 4.2). It has a low natural abundance and can exist

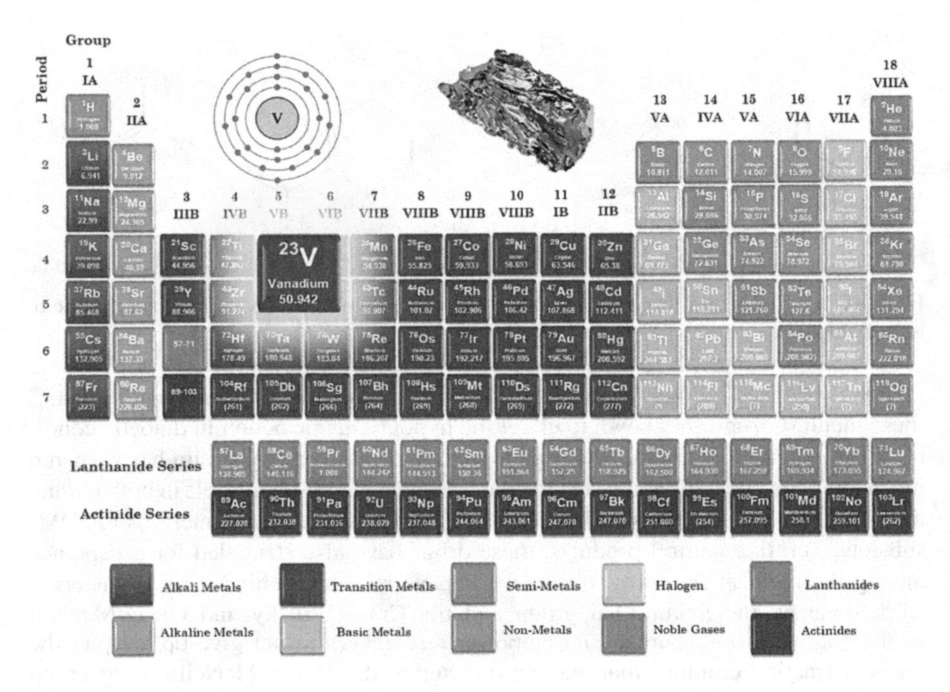

FIGURE 4.2 Periodic table highlighting the position, physical appearance, and atomic properties of vanadium.

in several oxidation states. However, once artificially isolated, the subsequent formation of an oxide layer (passivation) somewhat stabilizes the free metal against further reactivity and allows for its use in various applications.

In vitro and *in vivo* studies have demonstrated the potential of many vanadium compounds as insulin-mimetic (or insulin-enhancing) agents. Of particular interest was the observation that among those demonstrated to be effective in normalizing serum glucose and fatty acid levels in streptozotocin (STZ)-induced diabetics rat were complexes bearing ligands that were obtained from or modeled after natural sources, namely: bis(maltolato)oxovanadium(IV) (BMOV) (Figure 4.3a), bis(ethylmaltolato) oxovanadium(IV) (BEOV) (Figure 4.3b), diaqua(dipicolinato)oxovanadium(IV) (Figure 4.3c), and bis(allixinato)oxovanadium(IV) (Figure 4.3d) (D. C. Crans 2005, Adachi, Yoshida et al. 2006). Despite many attempts to explain how exactly vanadium compounds act in biological systems, only vague explanations persist. The problem is exacerbated by the fact that observations in tissue cultures do not always accurately reflect responses in animals. The enigma is rooted in their inconclusive mechanisms of action, uncertain dosage regimes, and speculated cytotoxicities (Thompson and Orvig 2006). Although vanadium has been taunted for its promiscuity, when it comes to intracellular protein binding, thus far it has been confirmed to contribute to dual ameliorative actions in the insulin transduction pathway. In the first mechanism (Mechanism 1), the vanadate ion directly binds to and activates the insulin receptor tyrosine kinase (IRTK) (Figure 4.4), thereby bypassing the need for the triggering effect of insulin. In its second mode of action (Mechanism 2),

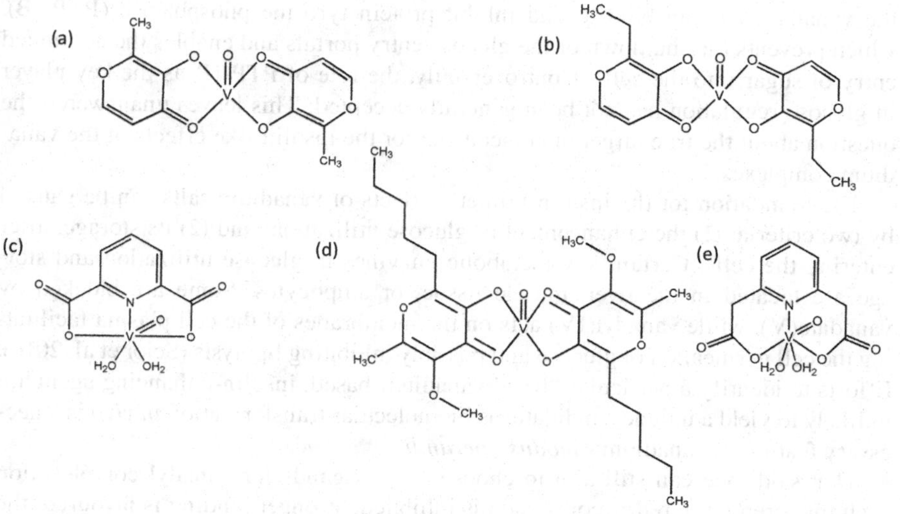

FIGURE 4.3 Structural formulae of vanadium complexes: (a) bis(maltolato) oxovanadium(IV): BMOV; (b) bis(ethylmaltolato)oxovanadium(IV): BEOV; (c) diaqua(dipicolinato)oxovanadium(IV); (d) bis(allixinato)oxovanadium(IV); and (e) [V(IV) O(2,6-pyridine diacetatato)$(H_2O)_2$]: PDOV.

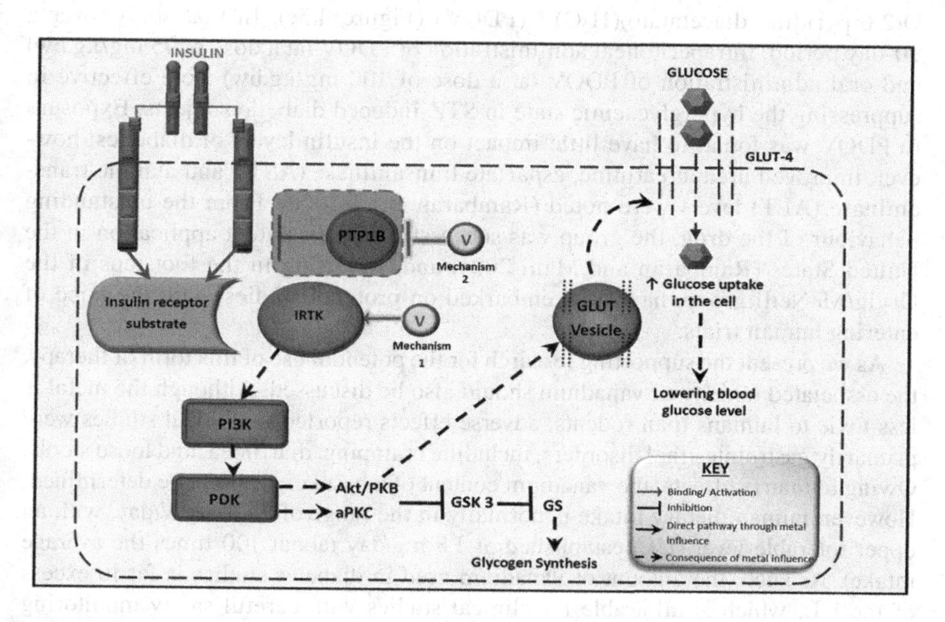

FIGURE 4.4 Mechanism of action of vanadium in the insulin transduction pathway.

the vanadate ion can bind to and inhibit protein tyrosine phosphatase (PTP-1B), which prevents the shutdown of the glucose entry portals and enables the continued entry of sugar into the cells. Controversially, the role of PTP-1B as the key player in glucose regulation has not been generally accepted. This leaves unanswered the question about the true target biomolecule(s) for the insulin-like effects of the vanadium complexes.

The foundation for the insulin-mimetic effects of vanadium salts can be gauged by two criteria: (1) the enhancement of glucose utilization; and (2) its storage, after entering the cells. Certain key metabolic enzymes in glucose utilization and storage are located in the liver, muscle tissue, or adipocytes. Some are blocked by vanadate(V), while vanadyl(IV) acts on the membranes of the cell plasma facilitating the cell permeation of glucose and possibly inhibiting lipolysis (Scior et al. 2016). Efforts to identify a particular "best" vanadium-based, insulin-enhancing agent are unlikely to yield a unique candidate, as biomolecular transformation *in vivo* is a necessary feature of vanadium's *modus operandi.*

That said, one can still aim to choose better ligands for vanadyl complexation such that premature redox conversion is inhibited, stronger binding is favoured, the safety of the dissociated ligand is assured, and synergistic effects are taken advantage of, where possible (Thompson and Orvig 2006).

As mentioned earlier, much research has been devoted to vanadium-centred coordination complexes, with one in particular making it as far as Phase 2 clinical trials (Thompson and Orvig 2006, Mohammadi et al. 2005, Thompson, Liboiron et al. 2003, Thompson, Lichter et al. 2009). Most recently, Rambaran et al. reported a unique glucose-lowering response to a novel complex [V(IV)O(2,6-pyridine diacetatato)(H_2O)$_2$] (PDOV) (Figure 4.3e). In that study, over a 90-day period, intraperitoneal administration of PDOV (at a dose of 75 mg/kg.bw) and oral administration of PDOV (at a dose of 100 mg/kg.bw) were effective in suppressing the hyperglycaemic state in STZ-induced diabetic subjects. Exposure to PDOV was found to have little impact on the insulin levels of diabetics; however, improved urea, creatinine, aspartate transaminase (AST), and alanine transaminase (ALT) levels were noted (Rambaran et al. 2020). From the outstanding behaviour of the drug, the group was successful in their patent application in the United States (Rambaran and Mani 2021) and, following in the footsteps of the Orvig/McNeill group, have now embarked on protocol studies, with the hope of entering human trials.

As we present the supporting research for the potential use of this form of therapy, the associated toxicity of vanadium should also be discussed. Although the metal is less toxic to humans than rodents, adverse effects reported in clinical studies were primarily gastrointestinal disorders, including cramping, diarrhoea, and loose stools. Owing to matrix effects, the vanadium content of a basal diet cannot be determined. However, human dietary intake is normally in the range of 5–20 µg V/day, with an upper tolerable level (UL) established at 1.8 mg/day (about 100 times the average intake). As such, the amount of vanadium used in diabetes studies is far in excess of the UL, which is allowable for clinical studies with careful safety monitoring (Vincent 2018).

4.3 CHROMIUM

Chromium is represented by the symbol Cr and atomic number 24 (Figure 4.5). It is debatably an essential element that is required for glucose and lipid metabolism and is further respected for its improvement of insulin sensitivity, by enhancing intracellular signalling. Despite its cons, supplemental trivalent chromium appears to be a useful tool in the world's fight against epidemic-like manifestations of metabolic syndromes, namely obesity and diabetes. Laboratory and clinical studies indicate that certain forms of trivalent chromium have various capabilities such as overcoming insulin resistance, ameliorating diabetes, suppressing free-radical formations, and decreasing systolic blood pressure (Peng and Yang 2015).

The intracellular action of chromium is still ungrounded, but studies have shown plausible evidence that, like vanadium, it binds to both IRTK and PTP-1B (Mechanisms 1 and 2) (R. A. Anderson 1998). Other studies have shown that it also assists in the promotion of insulin-induced glucose uptake via amplifying the activation of Akt (Mechanism 3) (Figure 4.6) (Peng and Yang 2015).

Chromium(III)picolinate (CrPic) (Figure 4.7) is the most extensively studied antidiabetic and anti-obese chromium complex. It has been reported that CrPic treatment is able to protect from hepatocellular injury and also shows enhanced Cr

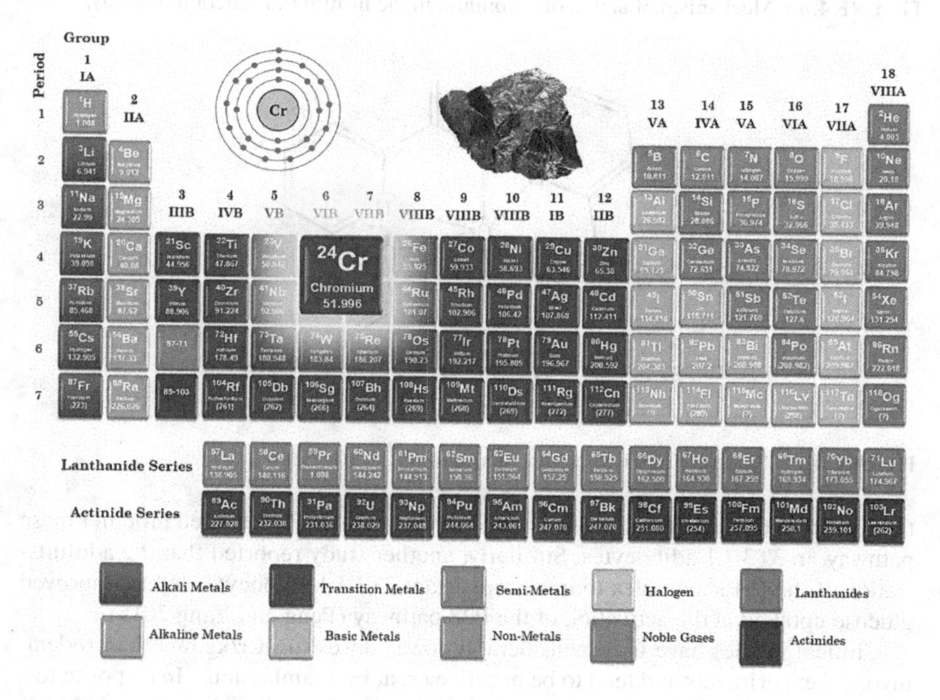

FIGURE 4.5 Periodic table highlighting the position, physical appearance, and atomic properties of chromium.

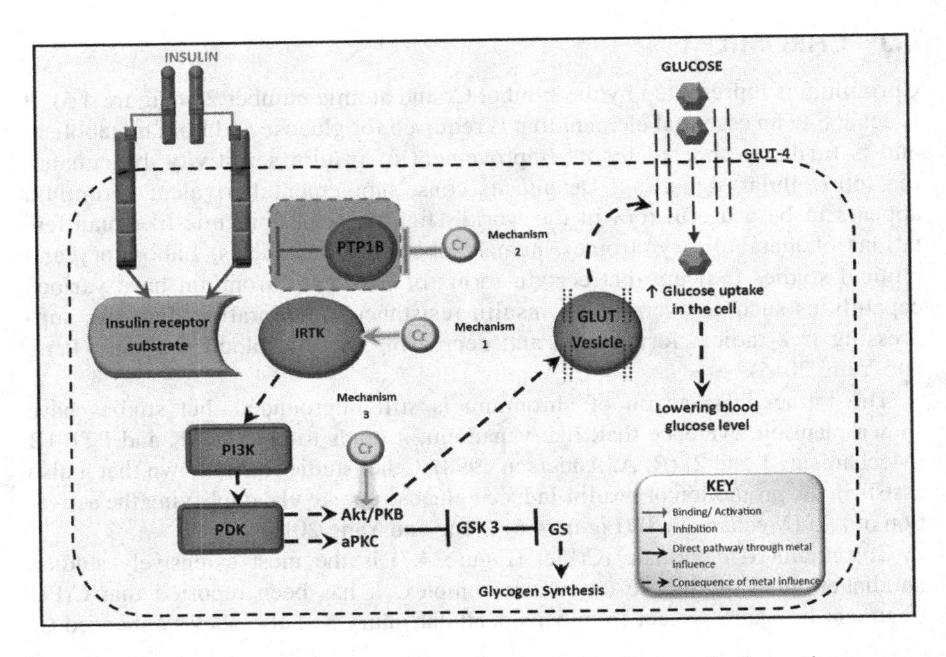

FIGURE 4.6 Mechanism of action of chromium in the insulin transduction pathway.

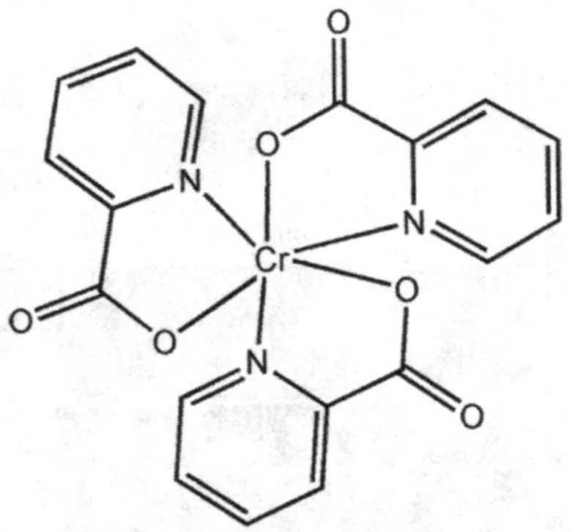

FIGURE 4.7 Structure of tris(picolinate)chromium(III) (CrPic).

translocation through the adenosine monophosphate (AMP)-activated protein kinase pathway, in 3T3-L1 adipocytes. Similarly, another study reported that the administration of the CrPic complex to insulin-resistant 3T3-L1 adipocytes led to improved glucose uptake via the activation of the P38 pathway (Peng and Yang 2015).

Clinical studies have used considerably lower doses (µg-Cr/kg.bw) than rodent-involved experiments and tend to be negative or, at best, ambiguous. In response to a request from a nutraceutical company, the FDA examined the relationship between Cr and insulin resistance, cardiovascular disease (CVD), T2DM, and other conditions related to elevated glucose levels. The FDA issued a letter of enforcement discretion allowing only one qualified health claim for the labelling of dietary supplements:

> One small study suggests that chromium picolinate may reduce the risk of insulin resistance, and therefore possibly may reduce the risk of T2DM. FDA concludes, however, that the existence of such a relationship between chromium picolinate and either insulin resistance or T2DM is highly uncertain.
>
> **(Vincent 2018)**

Not limiting its treatment to the classic cases of diabetes, we report here a study carried out by Janovic et al. In this study, chromium therapy was administered to a group of 30 women, who were diagnosed with gestational diabetes (20–24 gestational weeks). The study group was divided into three subgroups and given 0, 4, and 8 μg-Cr/kg.bw doses (in the form of CrPic) for 8 weeks. It was reported that chromium supplementation improved glucose intolerance and lowered hyperglycaemia in the treated patients. It was further noted that the group that received the 8 μg of supplement showed a higher degree of improvement relative to that of the group that received the 4 μg treatment. These results led the authors to conclude that "chromium picolinate supplementation may be an adjunctive therapy when dietary strategies are not sufficient to achieve normoglycaemia in women with gestational diabetes" (Jovanovic, Gutierrez and Peterson 1999).

It would be remiss of us if we did not also present the alleged shortcomings from the usage of CrPic. Questions arose about the safety in use of the complex as a dietary supplement, as Stearns et al. reported that the compound caused chromosomal cleavage in Chinese-hamster ovary (CHO) cells and was also mutagenic at the hypoxanthine phosphoribosyltransferase locus in CHO cells. However, the data was found to have been misrepresented as the studies used high, non-physiological concentrations of Cr (0.05–1 mM). Similarly, Bagchi et al. subsequently observed DNA fragmentation in another type of cultured cell treated with CrPic, although the Cr concentrations were also non-physiological. Apart from these extreme but debatable cases, there have been confirmed isolated incidents of the deleterious effects of the compound supplementation in humans, which include weight loss, anaemia, renal failure, and hypoglycaemia (Vincent 2003).

4.4 ZINC

Zinc is represented by the symbol Zn and atomic number 30 (Figure 4.8). Like chromium, it is considered an essential element and plays an important role in the maintenance of several tissue functions including the synthesis, storage, and release of insulin. The metal plays an important role in glucose metabolism and has been found to enhance the effectiveness of insulin *in vitro*. Further, it has been suggested that its deficiency may aggravate insulin resistance in non-insulin-dependent diabetes mellitus (NIDDM) (Devi et al. 2016).

The influence of zinc on the translocation of the glucose transporter type-4 (GLUT-4) vesicle to the plasma membrane has been proposed to be in the form of

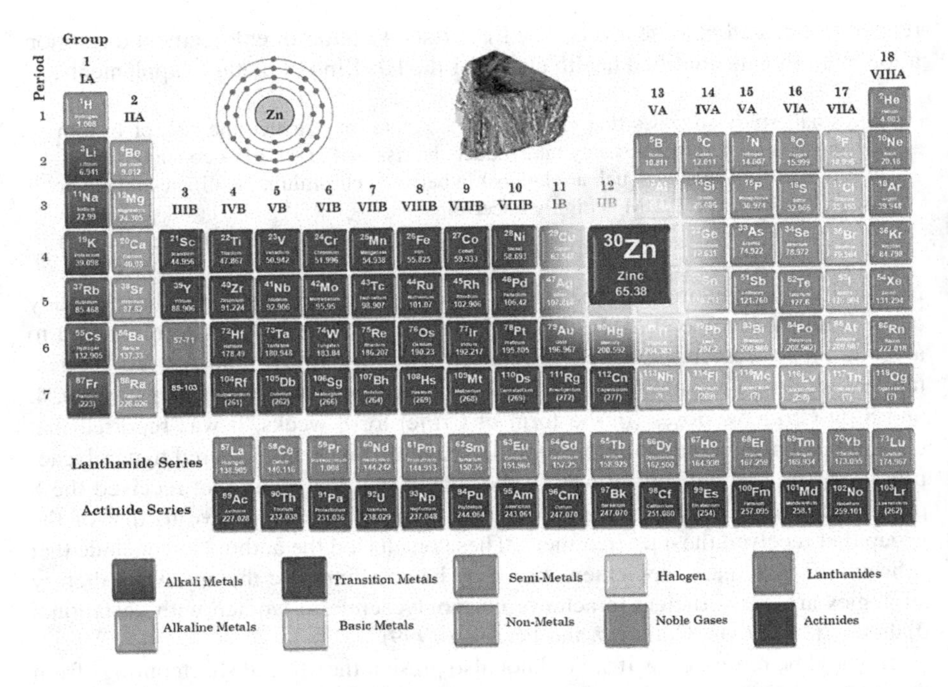

FIGURE 4.8 Periodic table highlighting the position, physical appearance, and atomic properties of zinc.

a post-insulin receptor mechanism. Supporting evidence has shown that the metal potentiates insulin-induced glucose transport and also increases glucose transport into cells by binding to key enzymes within the insulin signalling pathway.

The extent of zinc's interactions with the various intracellular proteins of the insulin transduction pathway is summarized in Figure 4.9. In the first instance, the metal ion influences the phosphorylation of the β-subunit of the insulin substrate receptor (INS-R) (Mechanism 1) and also the inhibition of the PTP-1B enzyme (Mechanism 2). Further to this, Akt is activated by zinc in a phosphatidylinositol 3-kinase (PI3K)-dependent way (Mechanism 3) and in a similar action to insulin, it also inhibits glycogen synthase kinase-3 (GSK-3) (Mechanism 4). Finally, the metal is also known to play a role in glucose transport, since it is part of the insulin-responsive aminopeptidase (IRAP) enzyme (a molecule required for the maintenance of normal GLUT levels) (Mechanism 5) (Jansen, Karges and Rink 2009).

Generally, it is accepted that upon binding with organic compounds, the associated toxicity of metal ions tends to decrease, while their relative bioavailability tends to increase. Supportively, an *in vitro* study by Sakurai et al. presented a series of Zn(II) complexes with maltol, picolinic acid, and amino acid ligands that had higher insulin-mimetic activity than zinc sulphate. In addition, the supplementation of KK-Ay mice (a T2DM model subject) with the Zn(II) complexes was found to maintain normoglycaemic levels (Ueda et al. 2002).

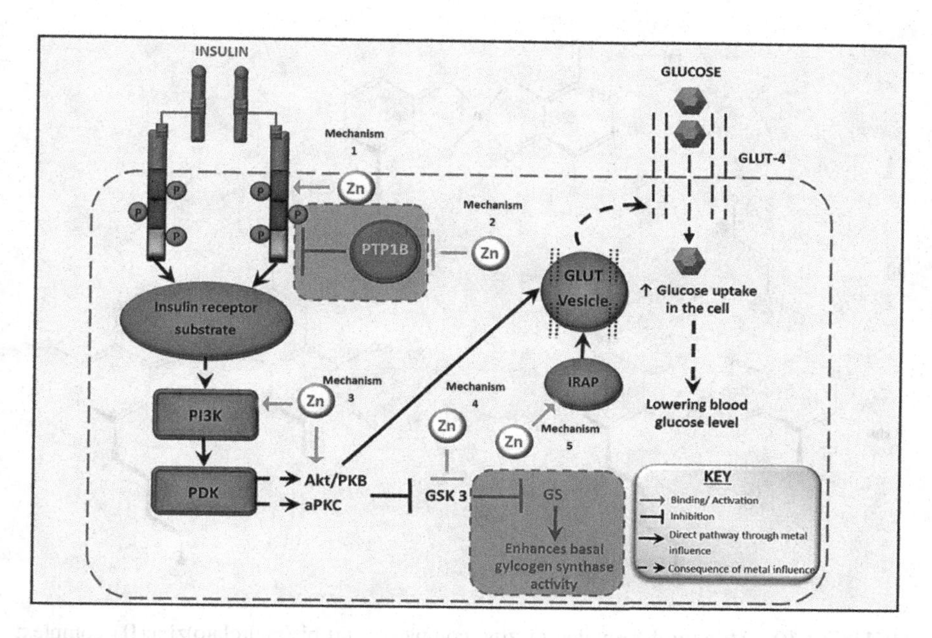

FIGURE 4.9 Mechanism of action of zinc in the insulin transduction pathway.

In another study, the Sakurai group used the potent complex: bis(maltolato)zinc(II) (Figure 4.10a) as the lead compound in their investigations of the structure–activity relationships (SARs) of its family of complexes. The *in vitro* insulin-mimetic activities of these complexes were determined by the inhibition of free fatty acid release and the enhancement of glucose uptake in isolated rat adipocytes treated with epinephrine. The group reported that the new Zn(II) complex, $Zn(alx)_2$ (Figure 4.10b), which was created from the allixin isolate (from garlic) exhibited the highest insulin-mimetic activity among all the complexes analyzed (Adachi, Yoshida et al. 2004).

Continuing with their investigations, Sakurai then showed that further modification of the "$Zn(alx)_2$" complex led to another novel Zn(II) compound, Zn(II)–thioallixin-N-methyl ($Zn(tanm)_2$) (Figure 4.10c). The results from this study showed that daily oral administration of $Zn(tanm)_2$ for 4 weeks in KKAy mice was able to significantly improve hyperglycaemia, glucose intolerance, insulin resistance, hyperleptinemia, obesity, and hypertension. In conclusion, the group proposed that their $Zn(tanm)_2$ complex could be used as a viable therapeutic option for obesity-linked T2DM and metabolic syndromes (Adachi, Yoshida et al. 2006, Adachi, Yoshikawa et al. 2006).

While caution taking zinc and its associated complexes is not as forewarned as vanadium and chromium, supplements should be taken responsibly and not excessively. According to the Office of Dietary Supplements (ODS), an excessive intake of zinc can lead to zinc toxicity, which can cause gastrointestinal discomfort, and when chronic, may also disrupt the balance of other chemicals in the body, including copper and iron (Galan 2019, National Institutes of Health 2021).

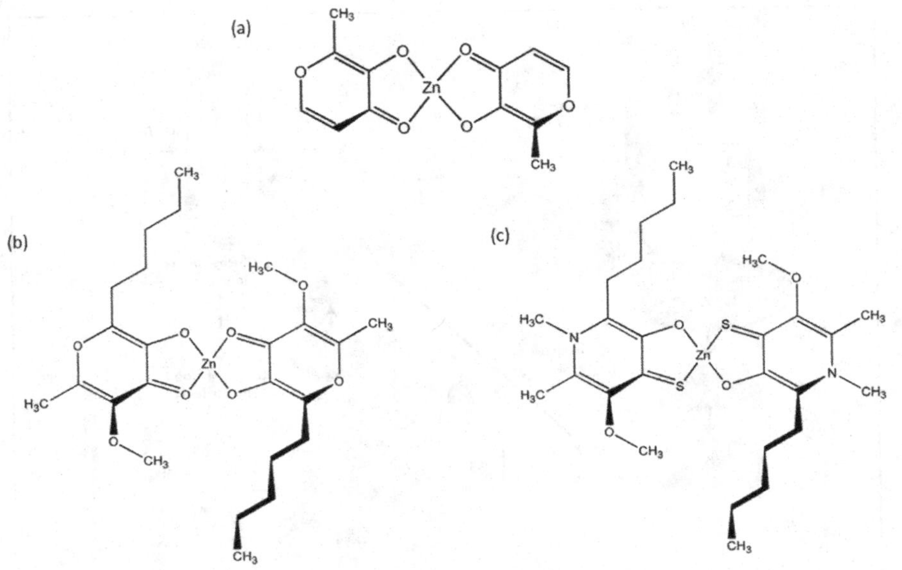

FIGURE 4.10 Structural formulae of zinc complexes: (a) bis(maltolato)zinc(II) complex; (b) bis(allixinato)Zn(II) complex: $Zn(alx)_2$; and (c) Zn(II)-thioallixin-N-methyl: $Zn(tanm)_2$.

4.5 COBALT

Cobalt is represented by the symbol Co and atomic number 30 (Figure 4.11). It is considered essential to the metabolism of all animals and is a key constituent of cobalamin (also known as vitamin B12) (Yamada 2013).

The literature has shown that under hypoxic conditions there is a responsive increase in GLUT-1 gene expression, which in turn leads to an increase in tissue glucose uptake. Interestingly, Beigi et al. reported that under normoxic conditions, the effect of hypoxia-induced GLUT-1 expression can be mimicked by the administration of Co in the form of cobalt chloride [$Co(II)Cl_2$] (Figure 4.12a) (Mechanism 1) (Ybarra et al. 1997).

This finding spurred the group's curiosity to investigate other glycaemic regulating mechanisms that can be stimulated by Co supplementation. In a following paper, they reported that the administration of 2 mM $CoCl_2$ in the drinking water of Sprague-Dawley rats led to a decrease in the production of phosphoenolpyruvate carboxykinase (PEPCK) mRNA (a rate-determining enzyme in the gluconeogenic pathway) (Figure 4.12b) (Mechanism 2). The decreased production of PEPCK had a consequential decrease in the production of hepatic glucose and thus presented itself as another mechanism in normoglycaemic regulation (Saker et al. 1998).

To further investigate the ameliorative effects of cobalt, McNeill et al. fed a population of both normal and diabetic rats a solution of 3.5 mM $CoCl_2$ (*ad libitum*) for 3 weeks, followed by a 4 mM solution for 4 weeks. Body weight and fluid consumption were monitored daily, while food intake was recorded twice every week. Prior to termination, an oral glucose tolerance test was performed on the animals. The

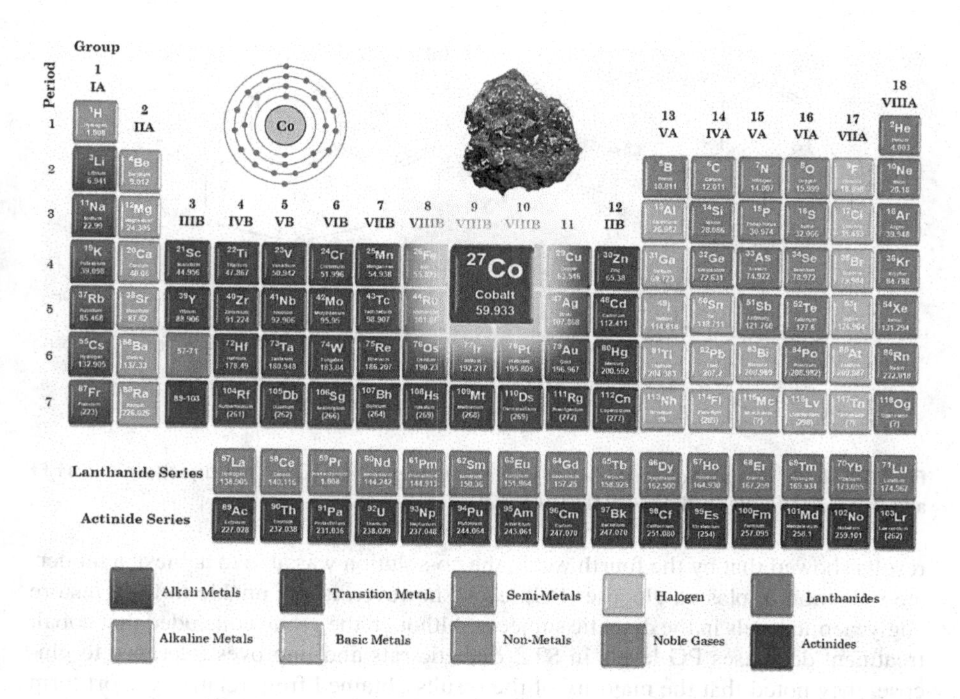

FIGURE 4.11 Periodic table highlighting the position, physical appearance, and atomic properties of cobalt.

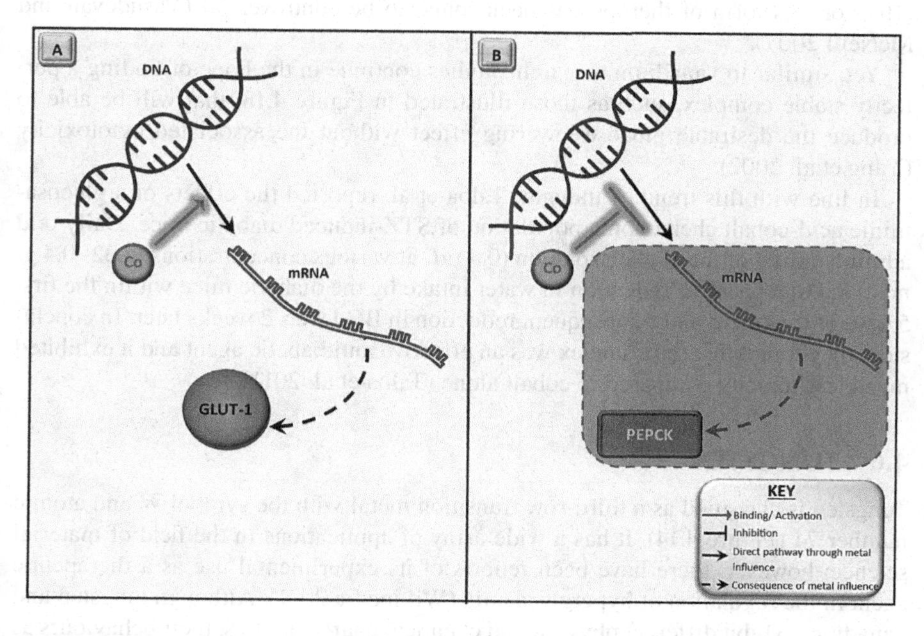

FIGURE 4.12 Illustration of (a) the role of Co(II) in the production of the GLUT-4 vesicle and (b) the inhibition of the production of PEPK, which forms part of the insulin transduction pathway.

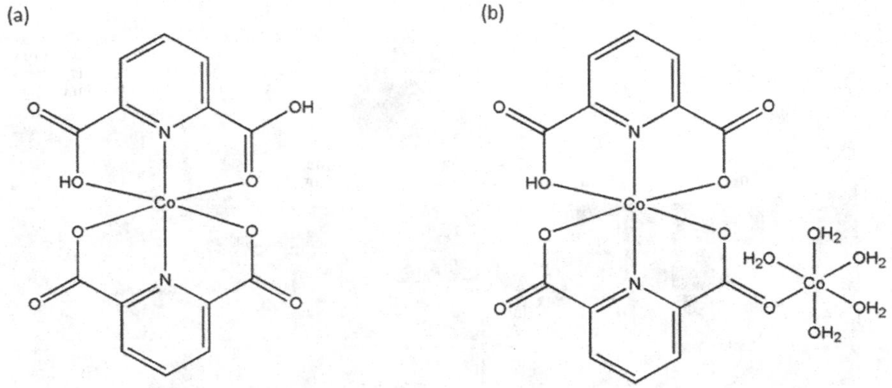

(a) (b)

FIGURE 4.13 Structural formulae of cobalt complexes: (a) [Co(II)(H_2dipic)(dipic)] $3H_2O$ and (b) [Co(II)(dipic)(μ-dipic)Co(II)$(H_2O)_5$] $2H_2O$.

results showed that by the fourth week, the co-solution was able to achieve a moderate reduction in plasma glucose (PG) levels; however, it was unable to fully restore euglycaemic levels in the diabetic subjects. Although the group concluded that cobalt treatment decreases PG levels in STZ-diabetic rats and improves tolerance to glucose, they noted that the majority of the results obtained from relatively short-term studies continued to reflect mediocre blood glucose (BG) attenuation. The fact that at higher concentrations the results have been shrouded by toxic events the beneficial effect of this form of therapy has been found to be controversial (Vasudevan and McNeill 2007).

Yet, similar to vanadium research, studies continue in the hope of finding a perfectly stable complex, such as those illustrated in Figure 4.13, that will be able to produce the desirable glucose-lowering effect without the associated cytotoxicity (Yang et al. 2002).

In line with this trend of thought, Talba et al. reported the effects of a glucosaminic acid-cobalt chelate on a population of STZ-induced diabetic mice. Daily oral administration of the chelate solution (0.4 mL at various concentrations [0.32–0.4 g/mL]) led to a peculiar reduction in water intake by the diabetic mice within the first 5 days of treatment and a subsequent reduction in BG levels 2 weeks later. In conclusion, they found that the complex was an effective antidiabetic agent and it exhibited much less toxicity compared to cobalt alone (Talba et al. 2011).

4.6 TUNGSTEN

Tungsten is classified as a third-row transition metal with the symbol W and atomic number 74 (Figure 4.14). It has a wide array of applications in the field of material science; however, there have been reports of its experimental use as a therapeutic agent in the regulation of hyperglycaemia (Wikipedia 2021). Although tungsten and vanadium exhibit different physical and chemical characteristics, their behaviours as insulin mimetic/enhancing agents bear great similarity to each other.

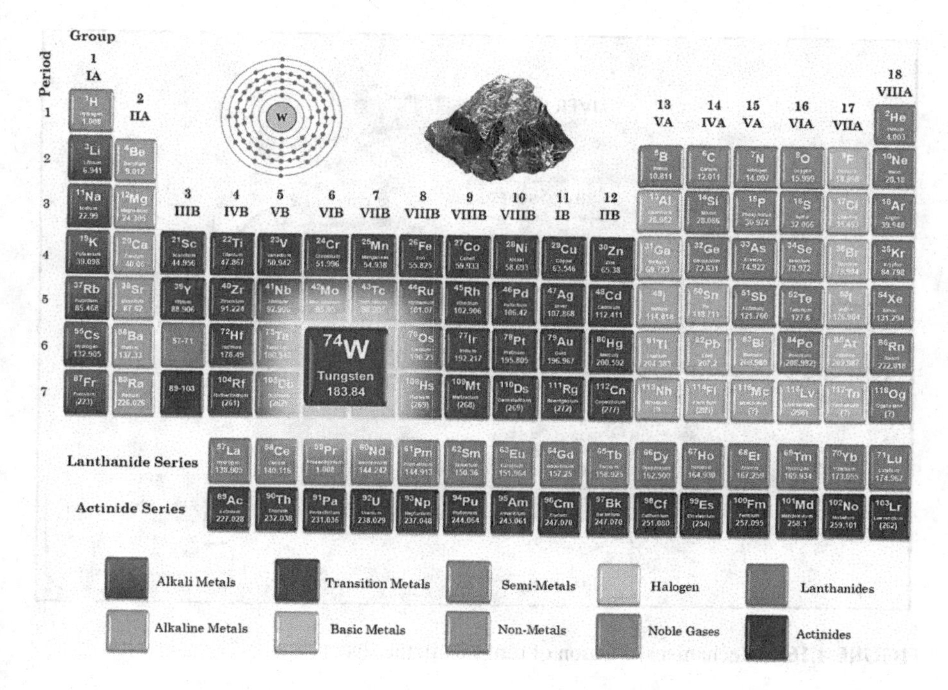

FIGURE 4.14 Periodic table highlighting the position, physical appearance, and atomic properties of tungsten.

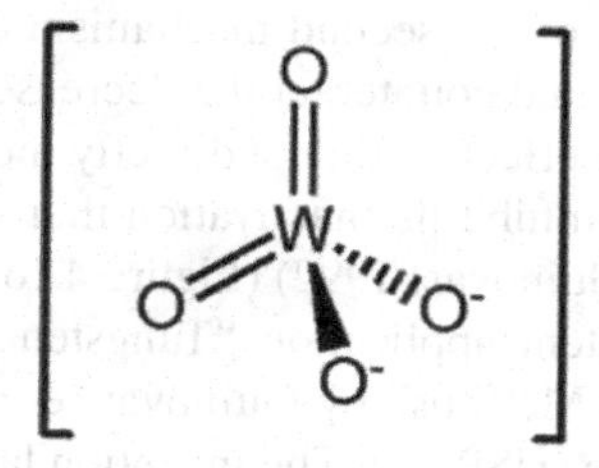

FIGURE 4.15 Structure of tungstate ion.

In recent years, a number of studies have demonstrated the antidiabetic effects of oral tungstate (WO_4^{2-}) (Figure 4.15) in diabetic animal models, and it has been touted as a more promising drug candidate than vanadium because of its relatively lower toxicity. Despite this slight edge over its competitor, the fear of the occurrence of cytotoxic events due to poor clearance is still a cause for concern (Domingo 2002).

Under diabetic conditions, there is a consequential increase in *L*-lactate (Jenei et al. 2019) and CO production (Paredi et al. 1999) and a decrease in glucose-6-phosphate levels (Aiston, Andersen and Agius 2003). Biological studies (using rat hepatocytes) have revealed that like vanadate, tungstate inactivates glycogen synthase (a

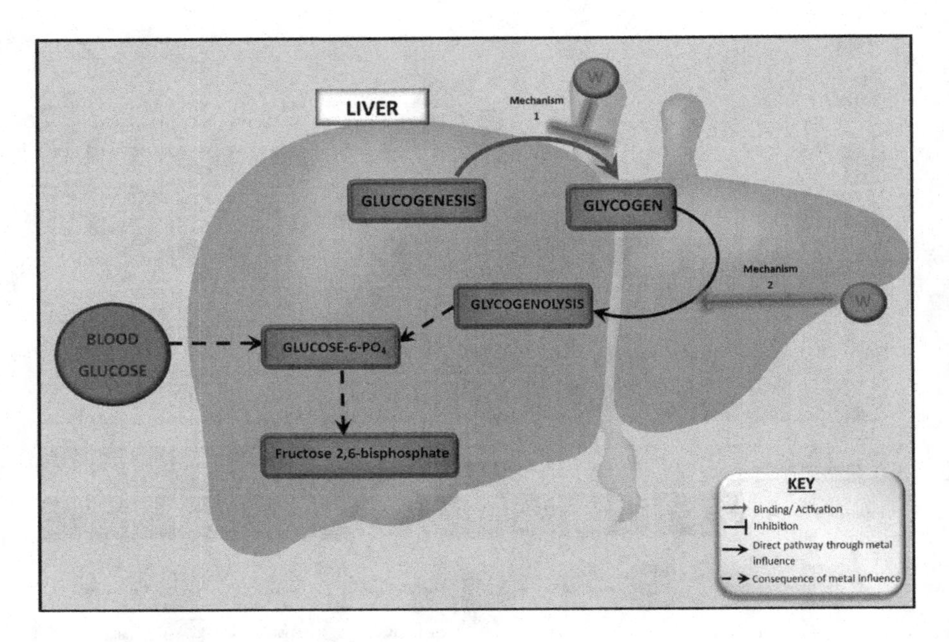

FIGURE 4.16 Mechanism of action of tungsten in the liver.

key enzyme in the conversion of glucose into glycogen) by a mechanism independent of Ca^{2+} (Mechanism 1). It is also seen to activate glycogen phosphorylase (an enzyme that catalyzes the rate-limiting step in glycogenolysis) by a Ca^{2+}-dependent pathway (Mechanism 2). The actions of the second mechanism consequently increase fructose 2,6-bisphosphate levels and counteract the decrease in this metabolite induced by glucagon. Although these effectors do not directly modify 6-phosphofructo-2-kinase activity, they partially inhibit its inactivation that was influenced by glucagon (Fillat, Rodriguez-Gil and Guinovart 1992) (Figure 4.16).

In January 1997, the patent application "Tungsten (VI) Compositions for the Oral Treatment of Diabetes Mellitus" by Guinovart et al. was approved by the US Patent and Trademark Office (USPTO). The invention laid claims to formulations of a tungstate-based, insulin-mimetic compound that was capable of attenuating hyperglycaemic levels in mammalian models through oral administration. Unfortunately, no further records were found on the advancement of the invention to the stage of human clinical trials, perhaps due to the looming fear of its toxicity (Agency for Toxic Substances and Disease Registry, Division of Toxicology and Human Health Sciences 2005).

Despite the understandable fears surrounding its use, studies on the toxic effects of tungstate remain very scant. It is therefore imperative that due to the continued interest in tungstate as a potential DM therapeutant, greater focus should be placed on the *in vitro* and *in vivo* evaluation of its toxicity (Guinovart Cirera, Barbera Lluis and Rodriguez-Gil 1997).

4.7 MOLYBDENUM

Molybdenum is classified as a second-row transition metal with the symbol Mo and atomic number 42 (Figure 4.17). It is an essential mineral that is found in high concentrations in legumes, grains, and organ meats, and is considered essential for human life (Rowles 2017). Like the tungstate ion, molybdate is a compound containing an oxo-anion with Mo in its highest oxidation state of 6 (Figure 4.18).

The insulin-like effects of Mo were studied in rat adipocytes and compared to the actions of vanadium. Other than being less potent, the molybdate's activities resembled those of the vanadate ion, in stimulating lipogenesis, activating

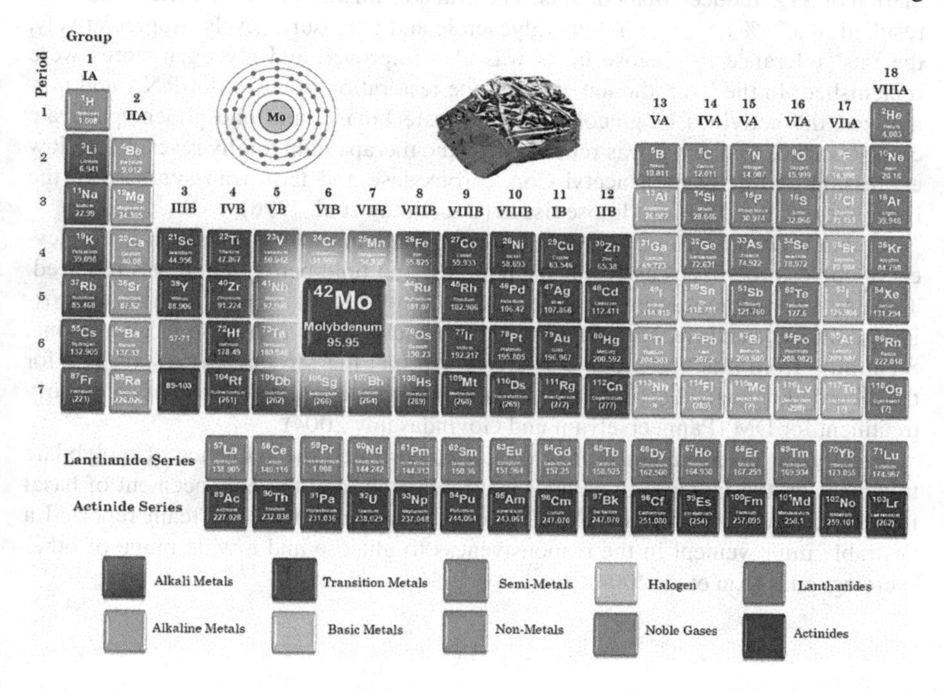

FIGURE 4.17 Periodic table highlighting the position, physical appearance, and atomic properties of molybdenum.

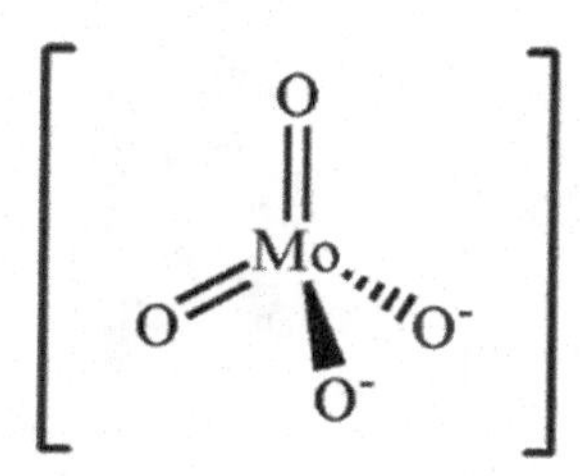

FIGURE 4.18 Structure of molybdate.

glucose oxidation, enhancing the rate of hexose uptake, and inhibiting lipolysis (Li et al. 1995).

Recently, evidence has been presented that Mo also influences carbohydrate metabolism *in vitro*. The metallo-oxo anion was observed to stimulate both glycolysis and accelerated glycogen degradation in isolated hepatocytes and was also seen to influence increased insulin receptor autophosphorylation, phosphorylation of insulin receptor 160-kDa substrate, and augmented glucose transport in adipocytes (Ozcelilkay et al. 1996).

A study by Ozcelilkay et al. reported the insulin-like action of molybdate in a population of STZ-induced diabetic rats. The oral administration of the metal to the rats resulted in a 75% decrease in hyperglycaemic and glucosuric levels. Appreciatively, the rats' tolerance to glucose loads was also improved and glycogen stores were replenished. In the liver, the ion effected the restoration of blunted mRNA and normalized the activities of glucokinase, pyruvate kinase, and phosphoenolpyruvate carboxykinase. Finally, it was reported that the therapeutant totally reversed the low expression and activity of acetyl-CoA carboxylase and fatty acid synthase in the liver, but not in the white adipose tissue (Ozcelilkay et al. 1996).

In another study (using molybdate-treated diabetic rats), a reduction in the levels of cholesterol, triglycerides, phospholipids, and lipid peroxidation was reported. There was also an observed increase in the activity of antioxidants such as superoxide dismutase, catalase, and glutathione peroxidase, and a reduction in glutathione. Stemming from this, the group was convinced that the Mo ion was responsible for the recorded biological activity and suggested its use as a form of prevention or early treatment for DM (Panneerselvam and Govindasamy 2004).

Finally, an investigation by Liu et al. demonstrated the influence of molybdate treatment on the increase in cellular-insulin content and the enhancement of basal insulin release, by clonal BRIN BD11 cells. Further to this, the team reported a desirable improvement in the responsiveness to glucose and a wide range of other secretagogues (Liu et al. 2004).

Conclusion

Board-approved antidiabetic drugs continue to be the most popular form of oral therapy for diabetes mellitus (DM), owing to the protocolled number of pharmacokinetic and pharmacodynamic studies carried out on them. However, influenced by both custom and cost, ethnopharmacological therapies also remain equally popular among patients with the disease. Apart from these allopathic and Ayurvedic medicines, ongoing research into insulin-mimetic coordination complexes has continued to generate great interest due to their respective improved efficacies and lower cytotoxicities. Although these medications may be controversially better than their Food and Drug Administration (FDA)-approved counterparts, their respective safety and efficacy standards need to be further evaluated by well-designed, controlled clinical studies.

Bibliography

Abbas, Nahid, Mashael H Al-Sueaadi, Ahlaam Rasheed, and Eiman S Ahmed. 2018. "Study of antidiabetic effect of lemongrass (*Cymbopogon citratus*) aqueous roots and flower extracts on albino mice." *International Journal of Pharmaceutical Sciences and Research* 9 (8): 3552–3555.

Abbas, Samir Y, Wahid M Basyouni, Khairy A M El-Bayouk, Wael M Tohamy, Hanan F Aly, Azza Arafa, and Mahmoud S Soliman. 2017. "New biguanides as anti-diabetic agents part I: Synthesis and evaluation of 1-substituted biguanide derivatives as anti-diabetic agents of type II diabetes insulin resistant." *Drug Research* 67 (10): 557–563. doi:10.1055/s-0043-102692.

Abdulazeez, Mansurah A, Kassim Ibrahim, Kenpia Bulus, Hope B Babvoshia, and Yusuf Abdullahi. 2013. "Effect of combined use of *Ocimum gratissimum* and *Vernonia amygdalina* extract on the activity of angiotensin converting enzyme, hypolipidemic and antioxidant parameters in streptozotocin-induced diabetic rats." *African Journal of Biochemistry Research* 7 (9): 165–173. doi:10.5897/AJBR12.091.

Abozid, Medhat M, Hanaa S M A El-Rahman, and Salama M Mohamed. 2018. "Evaluation of the potential anti-diabetic effect of *Apium graveolens* and *Brassica oleracea* extracts in alloxan induced diabetic rats." *International Journal of Pharmaceutical Sciences Review and Research* 49 (2): 39–44.

Abuelizz, Hatem A, El H Anouar, Rohaya Ahmad, Nor I I N Azman, Mohamed Marzouk, and Rashad Al-Salahi. 2019. "Triazoloquinazolines as a new class of potent α-glucosidase inhibitors: In vitro evaluation and docking study." *PLOS ONE* 14 (8): 1–13. doi:10.1371/journal.pone.0220379.

Adachi, Yusuke, Jiro Yoshida, Yukihiro Kodera, Akira Katoh, Jitsuya Takada, and Hiromu Sakurai. 2006. "Bis(allixinato)oxovanadium(IV) complex is a potent antidiabetic agent: Studies on structure–activity relationship for a series of hydroxypyrone–Vanadium complexes." *Journal of Medicinal Chemistry* 11 (3251–3256): 49.

Adachi, Yusuke, Jiro Yoshida, Yukihiro Kodera, Akira Kato, Yutaka Yoshikawa, Yoshitane Kojima, and Hiromu Sakurai. 2004. "A new insulin-mimetic bis(allixinato)zinc(II) complex: Structure-activity relationship of zinc(II) complexes." *Journal of Biological Inorganic Chemistry* 9 (7): 885–893. doi:10.1007/s00775-004-0590-8.

Adachi, Yusuke, Jiro Yoshida, Yukihiro Kodera, Tamas Kiss, Tamas Jakusch, Eva A Enyedy, Yutaka Yoshikawa, and Hiromu Sakurai. 2006. "Oral administration of a zinc complex improves type 2 diabetes and metabolic syndromes." *Biochemical and Biophysical Research Communications* 351 (1): 165–170. doi:10.1016/j.bbrc.2006.10.014.

Adachi, Yusuke, Yutaka Yoshikawa, Jiro Yoshida, Yukihiro Kodera, Akira Katoh, Yoshitane Kojima, and Hiromu Sakurai. 2006. "Zinc(II) complexes with allixin-derivatives as oral therapeutics for type 2 diabetes." *Biomedical Research on Trace Elements* 17 (1): 17–24.

Adedapo, Adeolu, Sunday Ofuegbe, and Oluwafemi Oguntibeju. 2014. "The antidiabetic activities of the aqueous leaf extract of *Phyllanthus amarus* in some laboratory animals." Chap. 5 in *Antioxidant-Antidiabetic Agents and Human Health*, edited by Oluwafemi Oguntibeju, 115–137. IntechOpen. doi:10.5772/57030.

Adedosu, O T, R A Jimoh, M A Saraki, and J A Badmu. 2018. "Ethanol leaves extract of *Mangifera indica* (L.) exhibits protective, antioxidative, and antidiabetic effects in rats." *Asian Pacific Journal of Health Sciences* 5 (1): 182–188. doi:10.21276/apjhs.2018.5.1.42.

Ademuyiwa, Adegbegi J, Ogunyemi Y Olamide, and Oyebiyi O Oluwatosin. 2015. "The effects of *Cymbopogon citratus* (lemon grass) on the blood sugar level, lipid profiles and hormonal profiles of wistar albino rats." *Merit Research Journal of Medicine and Medical Sciences* 3 (6): 210–216.

Adeneye, A A, Olufemi O Amole, and Adejuwon K Adeneye. 2006. "Hypoglycemic and hypocholesterolemic activities of the aqueous leaf and seed extract of *Phyllanthus amarus* in mice." *Fitoterapia* 77 (7–8): 511–514. doi:10.1016/j.fitote.2006.05.030.

Adeneye, Adejuwon A. 2012. "The leaf and seed aqueous extract of *Phyllanthus amarus* improves insulin resistance diabetes in experimental animal studies." *Journal of Ethnopharmacology* 144: 705–711. doi:10.1016/j.jep.2012.10.017.

Adeneye, Adejuwon A, and Esther O Agbaje. 2007. "Hypoglycemic and hypolipidemic effects of fresh leaf aqueous extract of *Cymbopogon citratus* Stapf. in rats." *Journal of Ethnopharmacology* 112: 440–444. doi:10.1016/j.jep.2007.03.034.

Adewole, Stephen O, and John A O Ojewole. 2009. "Protective effects of *Annona muricata* Linn. (Annonaceae) leaf aqueous extract on serum lipid profiles and oxidative stress in hepatocytes of streptozotocin-treated diabetic rats." *African Journal of Traditional, Complementary and Alternative Medicines* 6 (1): 30–41. doi:10.4314/ajtcam.v6i1.57071.

Adeyemi, David O, Omobola A Komolafe, Olarinde S Adewole, Efere M Obuotor, and Thomas K Adenowo. 2009. "Anti hyperglycemic activities of *Annona muricata* (Linn)." *African Journal of Traditional, Complementary and Alternative Medicines* 6 (1): 62–69.

Agawane, Sachin B, Vidya S Gupta, Mahesh J Kulkarni, Asish K Bhattacharya, and Santosh S Koratkar. 2017. "Chemo-biological evaluation of antidiabetic activity of *Mentha arvensis* L. and its role in inhibition of advanced glycation end products." *Journal of Ayurveda and Integrative Medicine* 1–5. doi:10.1016/j.jaim.2017.07.003.

Agbai, E O, C J Njoku, C O Nwanegwo, and A C Nwafor. 2015. "Effect of aqueous extract of *Annona muricata* seed on atherogenicity in streptozotocin-induced diabetic rats." *African Journal of Pharmacy and Pharmacology* 9 (30): 745–755. doi:10.5897/AJPP2015.4389.

Agbai, E O, P P E Mounmbegna, C J Njoku, C O Nwanegwo, G A Awemu, and S C Iwuji. 2015. "Effect of *Annona muricata* seed extract on blood glucose, total and differential white cell count after repeated exposure to clozapine." *Research in Neuroscience* 4 (1): 10–15. doi:10.5923/j.neuroscience.20150401.02.

Agency for Toxic Substances and Disease Registry, Division of Toxicology and Human Health Sciences. 2005. *Toxicological Profile for Tungsten*. Atlanta.

Aggarwal, Bharat B, Chitra Sundaram, Nikita Malani, and Haruyo Ichikawa. 2007. "Curcumin: The Indian solid gold." *Advances in Experimental Medicine and Biology* 595: 1–75. doi:10.1007/978-0-387-46401-5_1.

Agu, Kinsgley C, Nkeiruka Eluehike, Reuben O Ofeimun, Deborah Abile, Godwin Ideho, Marianna O Olukemi, Priscilla O Onose, and Olusola O Elekofehinti. 2019. "Possible anti-diabetic potentials of *Annona muricata* (soursop): Inhibition of α-amylase and α-glucosidase activities." *Clinical Phytoscience* 5 (21): 1–13. doi:10.1186/s40816-019-0116-0.

Aguiyi, John C, C I Obi, S S Gang, and Augustine C Igweh. 2000. "Hypoglycaemic activity of *Ocimum gratissimum* in rats." *Fitoterapia* 71: 444–446. doi:10.1016/s0367-326x(00)00143-x.

Agunbiade, Olabode S, O M Ojezele, J O Ojezele, and A Y Ajayi. 2012. "Hypoglycaemic activity of *Commelina africana* and *Ageratum conyzoides* in relation to their mineral composition." *African Health Sciences* 12 (2): 198–203. doi:10.4314/ahs.v12i2.19.

Ahmad, Hafsa, Sakshi Sehgal, Anurag Mishra, and Rajiv Gupta. 2012. "*Mimosa pudica* L. (Laajvanti): An overview." *Pharmacognosy Reviews* 6 (12): 115–124. Accessed August 13, 2020. doi:10.4103/0973-7847.99945.

Ahmad, Jamil, Imran Khan, and Renald Blundell. 2019. "*Moringa oleifera* and glycemic control: A review of current evidence and possible mechanisms." *Phtyotherapy Research* 33 (11): 2841–2848. doi:10.1002/ptr.6473.

Ahmed, Mohammed F, Syed M Kazim, Syed S Ghori, Syeda S Mehjabeen, Shaik R Ahmed, Shaik M Ali, and Mohammed Ibrahim. 2010. "Antidiabetic activity of *Vinca rosea* extracts in alloxan-induced diabetic rats." *International Journal of Endocrinology* 2010 (1): 1–7.

Airaodion, Augustine I, Emmanuel O Ogbuagu, John A Ekenjoku, Uloaku Ogbuagu, and Victor N Okoroukwu. 2019. "Antidiabetic effect of ethanolic extract of *Carica papaya* leaves in alloxan-induced diabetic rats." *American Journal of Biomedical Science & Research* 5 (3): 227–234. doi:10.34297/AJBSR.2019.05.000917.

Aiston, Susan, Birgitte Andersen, and Loranne Agius. 2003. "Glucose 6-phosphate regulates hepatic glycogenolysis through inactivation of phosphorylase." *Diabetes* 52 (6): 1333–1339.

Akhila, S, N Aleykutty, and P Manju. 2012. "Docking studies on *Peperomia pellucida* as antidiabetic drug." *International Journal of Pharmacy and Pharmaceutical Sciences* 4 (supp 4): 76–77.

Akhlaghi, Masoumeh, and Sahar Foshati. 2017. "Bioavailability and metabolism of flavonoids: A review." *International Journal of Nutrition Sciences* 2 (4): 180–184.

Al Rawi, Sara N, Amal Khidir, Maha S Elnashar, Huda A Abdelrahim, Amal K Killawi, Maya M Hammoud, and Michael D Fetters. 2017. "Traditional Arabic & Islamic medicine: Validation and empirical assessment of a conceptual model in Qatar." *BMC Complementary and Alternative Medicine* 17 (157): 1–10.

Al-Amin, Zainab M, Martha Thomson, Khaled K Al-Qattan, Riitta Peltonen-Shalaby, and Muslim Ali. 2006. "Anti-diabetic and hypolipidaemic properties of ginger (*Zingiber officinale*) in streptozotocin-induced diabetic rats." *British Journal of Nutrition* 96 (4): 660–666. doi:10.1079/bjn20061849.

Aligita, Widhya, Elis Susilawati, Harini Septiani, and Raihana Atsil. 2018. "Antidiabetic activity of coriander (*Coriandrum sativum* L) leaves' ethanolic extract." *International Journal of Pharmaceutical and Phytopharmacological Research* 8 (2): 59–63.

AlMalaak, Maha K, Nasir A A Almansour, and Zahida M Hussein. 2018. "Antihyperglycemic effect of n-butanol extract of celery (*Apium graveolens*) seeds and expression level of pancreatic, placental and fetal Sox17, Pax6, Ins1, Ins2 and glucagon genes in STZ-induced." *AL-Qadisiya Medical Journal* 14 (25): 124–146. doi:10.13140/RG.2.2.12322.58562.

Al-Sa'aidi, Jabbar A A, and Basim A K Al-Shihmani. 2013. "Antihyperglycaemic and pancreatic regenerative effect of n-butanol extract of celery (*Apium graveolens*) seed in STZ-induced diabetic male rats." *Suez Canal Veterinary Medical Journal* 18 (1): 71–85.

Al-Snafi, Ali E. 2017. "Pharmacology and therapeutic potential of *Euphorbia hirta* (Syn: *Euphorbia pilulifera*) - A review." *IOSR Journal of Pharmacy* 7 (3): 7–20. doi:10.9790/3013-0703010720.

Amaliah, Ulfah N, Eva Johannes, Munif S Hassan, and Elis Tambaru. 2019. "The use extract of siam leaf *Eupatorium odoratum* L. as alternative material in lowering blood glucose." *International Journal of Applied Biology* 3 (1): 15–23.

Amarathunga, A A M D D N, and S U Kankanamge. 2017. "Review on pharmacognostic, phytochemical and ethnopharmacological findings of *Peperomia pellucida* (L.) Kunth: Pepper elder." *International Research Journal of Pharmacy* 8 (11): 16–23. doi:10.7897/2230-8407.0811211.

American Diabetes Association. 2013. "Standards of medical care in diabetes." *Diabetes Care* 36 (Suppl 1) (1): S11–S66. doi:10.2337/dc13-S011.

Anderson, Leigh Ann, Sanjai Sinha, Kaci Durbin, Sophia Entringer, Judith Stewart, Philip Thornton, Carmen Fookes, et al. 2021. *Tolbutamide Side Effects.* September 29. https://www.drugs.com/sfx/tolbutamide-side-effects.html.

Anderson, Leigh Ann, Sanjai Sinha, Kaci Durbin, Sophia Entringer, Judith Stewart, Philip Thornton, Carmen Fookes, et al. 2022. *TOLINAZE.* 03 10. https://www.drugs.com/sfx/tolazamide-side-effects.html.

Anderson, Richard A. 1998. "Chromium, glucose intolerance and diabetes." *Journal of the American College of Nutrition* 17 (6): 548–555. doi:10.1080/07315724.1998.10718802.

Andrade-Cetto, Adolfo, and René Cárdenas-Vázquez. 2010. "Gluconeogenesis inhibition and phytochemical composition of two Cecropia species." *Journal of Ethnopharmacology* 130 (1): 93–97. doi:10.1016/j.jep.2010.04.016.

Andrade-Cetto, Adolfo, Christian A Cabello-Hernández, and Rene Cárdenas-Vázquez. 2015. "Alpha-glucosidase inhibiting activity of five Mexican plants used in the treatment of type 2 diabetes." *Pharmacologyonline* 1: 67–71. doi:10.1055/s-0033-1352350.

Andrade-Cetto, Adolfo, Cristina Revilla-Monsalve, and Wiedenfeld Wiedenfeld. 2007. "Hypoglycemic effect of *Tournefortia hirsutissima* L., on n-streptozotocin diabetic rats." *Journal of Ethnopharmacology* 112 (1): 96–100. doi:10.1016/j.jep.2007.02.020.

Andrade-Cetto, Adolfo, Elda C Cruz, Christian A Cabello-Hernández, and René Cárdenas-Vázquez. 2019. "Hypoglycemic activity of medicinal plants used among the cakchiquels in Guatemala for the treatment of type 2 diabetes." *Evidence-Based Complementary and Alternative Medicine* 2019, 1–7.

Andrade-Cetto, Adolfo, René Cárdenas, and Brenda Ramírez-Reyes. 2007. "Hypogylycemic effect of *Cecropia peltata* L. on N5-STZ type 2 diabetic rats." *Pharmacologyonline* 3: 203–210.

Andrews, Colleen M, Kevin Wyne, and James E Svenson. 2018. "The use of traditional and complementary medicine for diabetes in rural Guatemala." *Journal of Health Care for the Poor and Underserved* 29 (4): 1188–1208. doi:10.1353/hpu.2018.0092.

Angel, Jose, Kumar Sai Sailesh, and J K Mukkadan. 2013. "A study on anti-diabetic effect of peppermint in alloxan induced diabetic model of wistar rats." *Journal of Clinical and Biomedical Sciences* 3 (4): 177–181.

Antora, Raiya A, and Rabeta M Salleh. 2017. "Antihyperglycemic effect of *Ocimum* plants: A short review." *Asian Pacific Journal of Tropical Biomedicine* 7 (8): 755–759. doi:10.1016/j.apjtb.2017.07.010.

Aransiola, Elizabeth F, M O Daramola, Ezekiel O Iwalewa, A M Seluwa, and O O Olufowobi. 2014. "Anti-diabetic effect of *Bryophyllum pinnatum* leaves." *International Journal of Biotechnology and Bioengineering* 8 (1): 89–93.

Arenal, Amilcar, Leonardo Martín, Nestor M Castillo, Dainier de la Torre, Ubaldo Torres, and Reinaldo González. 2012. "Aqueous extract of *Ocimum tenuiflorum* decreases levels of blood glucose in induced hyperglycemic tilapia (*Oreochromis niloticus*)." *Asian Pacific Journal of Tropical Medicine* 5 (8): 634–637. doi:10.1016/S1995-7645(12)60130-8.

Arulselvan, Palanisamy, and Sorimuthu P Subramanian. 2007. "Beneficial effects of *Murraya koenigii* leaves on antioxidant defense system and ultra structural changes of pancreatic beta-cells in experimental diabetes in rats." *Chemico-Biological Interactions* 165 (2): 155–164. doi:10.1016/j.cbi.2006.10.014.

Arun, N, and N Nalini. 2002. "Efficacy of turmeric on blood sugar and polyol pathway in diabetic albino rats." *Plant Foods for Human Nutrition* 57 (1): 41–52. doi:10.1023/a:1013106527829.

Arya, Aditya, Mahmood A Abdullah, Batoul S Haerian, and Mustafa A Mohd. 2012. "Screening for hypoglycemic activity on the leaf extracts of nine medicinal plants: In-vivo evaluation." *E-Journal of Chemistry* 9 (3): 1196–1205.

Aschner, Pablo. 2017. "New IDF clinical practice recommendations for managing type 2 diabetes in primary care." *Diabetes Research and Clinical Practice* 132: 169–170.

Asomugha, Roseline N, P N Okafor, Ifeoma I Ijeh, Orish E Orisakwe, A L Asomugha, and J C Ndefo. 2013. "Toxicological evaluation of aqueous leaf extract of *Chromolaena odorata* in male wistar albino rats." *Journal of Applied Pharmaceutical Science* 3 (12): 89–92. doi:10.7324/JAPS.2013.31216.

AstraZeneca Pharmaceuticals. 2020. "FARXIGA." 05. Accessed 03 10, 2022. https://www.accessdata.fda.gov/drugsatfda_docs/label/2020/202293s020lbl.pdf.

Augusti, K T, and C G Sheela. 1996. "Antiperoxide effect of S-allyl cysteine sulfoxide, an insulin secretagogue, in diabetic rats." *Experientia* 52 (2): 115–120. doi:10.1007/BF01923354.

Augusti, K T, and M E Benaim. 1975. "Effect of essential oil of onion (allyl propyl disulphide) on blood glucose, free fatty acid and insulin levels of normal subjects." *Clinica Chimica Acta* 60 (1): 121–23. doi:10.1016/0009-8981(75)90190-4.

Ayyanar, Muniappan, and Pandurangan Subash-Babu. 2012. "*Syzygium cumini* (L.) Skeels: A review of its phytochemical constituents and traditional uses." *Asian Pacific Journal of Tropical Biomedicine* 2 (3): 240–246. doi:10.1016/S2221-1691(12)60050-1.

Azaizeh, Hassan, Bashar Saad, Edwin Cooper, and Omar Said. 2010. "Traditional Arabic and Islamic medicine, a re-emerging health aid." *Advance Access Publication* 7 (4): 419–424.

Babu, Sayyad S, Dasari B Madhuri, and Shaik L Ali. 2016. "A pharmacological review of *Urena lobata* plant." *Asian Journal of Pharmaceutical and Clinical Reseaarch* 9 (2): 20–22.

Bailey, Clifford J. 1992. "Biguanides and NIDDM." *Diabetes Care* 15 (6): 755–772. doi:10.2337/diacare.15.6.755.

Bailey, Clifford J, and Caroline Day. 1989. "Traditional plant medicines as treatments for diabetes." *Diabetes Care* 12 (8): 553–564. doi:10.2337/diacare.12.8.553.

Barbalho, Sandra M, Débora C Damasceno, Ana P M Spada, Vanessa Sellis da Silva, Karla A Martuchi, Marie Oshiiwa, Flavia M V F Machado, and Claudemir G Mendes. 2011. "Metabolic profile of offspring from diabetic wistar rats treated with *Mentha piperita* (peppermint)." *Evidence-Based Complementary and Alternative Medicine* 2011 (2): 1–6. doi:10.1155/2011/430237.

Bartimaeus, Ebirien-Agana S, Justice G Echeonwu, and Stella U Ken-Ezihuo. 2016. "The effect of *Cucumis sativus* (cucumber) on blood glucose concentration and blood pressure of apparently healthy individuals in port harcourt." *European Journal of Biomedical and Pharmaceutical Sciences* 3 (12): 108–114.

Basyouni, Wahid M, Samir Y Abbas, Mohamed F El Shehry, El-Bayouki Khairy A M, Hanan F Aly, Azza Arafa, and Mahmoud S Soliman. 2017. "New biguanides as anti-diabetic agents, part II: Synthesis and anti-diabetic properties evaluation of 1-arylamidebiguanide derivatives as agents of insulin resistant type II diabetes." *Archiv der Pharmazie* 350 (11): 1–9. doi:10.1002/ardp.201700183.

Bayani, Mahsan, Mahmood Ahmadi-hamedani, and Ashkan Jebelli Javan. 2017. "Study of hypoglycemic, hypocholesterolemic and antioxidant activities of Iranian *Mentha spicata* leaves aqueous extract in diabetic rats." *Iranian Journal of Pharmaceutical Research* 16 (Special Issue): 75–82.

Bayer Healthcare International Pharmaceuticals. 2012. "GLYSET." 08. Accessed 03 10, 2022. https://www.accessdata.fda.gov/drugsatfda_docs/label/2012/020682s010lbl.pdf.

Bayer Healthcare Pharmaceuticals. 2011. "PRECOSE." 03. Accessed 03 10, 2022. https://www.accessdata.fda.gov/drugsatfda_docs/label/2011/020482s024lbl.pdf.

Bhadoriya, Santosh S, Aditya Ganeshpurkar, Ravi P S Bhadoriya, Sanjeev K Sahu, and Jay R Patel. 2018. "Antidiabetic potential of polyphenolic-rich fraction of *Tamarindus indica* seed coat in alloxan-induced diabetic rats." *Journal of Basic and Clinical Physiology and Pharmcology* 29 (1): 37–45. doi:10.1515/jbcpp-2016-0193.

Bharti, Sudhanshu K, Amit Kumar, Om Prakash, Supriya Krishnan, and Ashok K Gupta. 2013. "Essential oil of *Cymbopogon citratus* against diabetes: Validation by in vivo experiments and computational studies." *Journal of Bioanalysis & Biomedicine* 5 (5): 194–203. doi:10.4172/1948-593X.1000098.

Bhaskar, Anusha, and Vidhya Gopalakrishnan. 2012. "Hypoglycemic and hypolipidemic activity of *Hibiscus rosa sinensis* Linn on streptozotocin–induced diabetic rats." *International Journal of Diabetes in Developing Countries* 32 (4): 214–218. doi:10.1007/s13410-012-0096-9.

Bhat, Menakshi, Sandeepkumar K Kothiwale, Amruta R Tirmale, Shobha Y Bhargava, and Bimba N Joshi. 2011. "Antidiabetic properties of *Azardiracta indica* and *Bougainvillea spectabilis*: In vivo studies in murine diabetes model." *Hindawi Publishing Corporation* 2011: 1–9. doi:10.1093/ecam/nep033.

Bhowmick, Debapriyo, Garima Bartariya, and Abhishek Kumar. 2018. "Qualitative analysis and a-amylase inhibition assay of aqueous foliar extract of *Lantana camara* (Linn)." *World Journal of Pharmaceutical Research* 7 (14): 920–925.

Bhowmik, Amrita, Liakot A Khan, Masfida Akhter, and Begum Rokeya. 2009. "Studies on the antidiabetic effects of *Mangifera indica* stem-barks and leaves on nondiabetic, type 1 and type 2 diabetic model rats." *Bangladesh Journal of Pharmacology* 4 (2): 110–114. doi:10.3329/bjp.v4i2.2488.

Billacura, Merell P, and Ian C T Alansado. 2017. "In vitro and in vivo hypoglycemic and colorimetric determination of glucose concentration of the different solvent extracts of *Crescentia cujete* Linn. fruit." *International Journal of Advanced and Applied Sciences* 4 (7): 21–28. doi:10.21833/ijaas.2017.07.005.

Biocon. 2022. "VOLICOSE." *Biocon*. 10 03. https://www.biocon.com/docs/prescribing_information/diabetology/volicose_pi.pdf.

Boaduo, N K K, D Katerere, J N Eloff, and V Naidoo. 2014. "Evaluation of six plant species used traditionally in the treatment and control of diabetes mellitus in South Africa using in vitro methods." *Pharmaceutical Biology* 52 (6): 756–761.

Boehringer Ingelheim Pharmaceuticals. 2022. "JARDIANCE." 02. Accessed 03 10, 2022. https://docs.boehringer-ingelheim.com/Prescribing%20Information/PIs/Jardiance/jardiance.pdf.

Boris, Azantsa K G, Djuikoo N Imelda, Kuikoua T Wilfried, Takuissu Guy, Judith L Ngondi, and Julius Oben. 2017. "Phytochemical screening and anti-diabetic evaluation of *Citrus sinensis* stem bark extracts." *International Journal of Biochemistry Research & Review* 17 (2): 1–13.

Bose, S, and G C Sepaha. 1956. "Clinical observations on the antidiabetic properties of *Pterocarpus marsupium* and *Eugenia jambolana*." *Journal of Indian Medical Association* 27 (11): 388–391.

Bristol-Myers. 1995. "Highlights of prescribing information: Glucophage and glucophage XR." Accessed 03 10, 2022. https://packageinserts.bms.com/pi/pi_glucophage.pdf.

Buchanan, Bob B, Wilhelm Gruissem, and Russell L Jones. 2000. *Biochemistry and Molecular Biology of Plants*. Rockville, MD: American Society of Plant Physiologists.

Campbell, Lance K, Danial E Baker, and R Keith Campbell. 2000. "Miglitol: Assessment of its role in the treatment of patients with diabetes mellitus." *The Annals of Pharmacotherapy* 34 (11): 1291–1301. doi:10.1345/aph.19269.

Campbell, R Keith. 1998. "Glimepiride: Role of a new sulfonylurea in the treatment of type 2 diabetes mellitus." *The Annals of Pharmacotherapy* 32 (10): 1044–1052. doi:10.1345/aph.17360.

Charbonnel, Bernard, Avraham Karasik, Ji Liu, Mei Wu, and Gary Meininger. 2006. "Efficacy and safety of the dipeptidyl peptidase-4 inhibitor sitagliptin added to ongoing metformin therapy in patients with type 2 diabetes inadequately controlled with metformin alone." *Diabetes Care* 29 (12): 2638–2643. doi:10.2337/dc06-0706.

Chattopadhyay, R R, S K Sarkar, S Ganguly, R N Banerjee, and T K Basu. 1991. "Hypoglycemic and antihyperglycemic effect of leaves of *Vinca rosea* Linn." *Indian Journal of Physiology and Pharmacology* 35 (3): 145–151.

Chougala, Mallikarjun B, Jamuna J Bhaskar, M G R Rajan, and Paramahans V Salimath. 2012. "Effect of curcumin and quercetin on lysosomal enzyme activities in streptozotocin-induced diabetic rats." *Clinical Nutrition* 31 (5): 749–755. doi:10.1016/j.clnu.2012.02.003.

Christensen, Hege, Anders Asberg, Aase-Britt Holmboe, and Knut J Berg. 2002. "Coadministration of grapefruit juice increases systemic exposure of diltiazem in healthy volunteers." *European Journal of Clinical Pharmacology* 58 (8): 515–520. doi:10.1007/s00228-002-0516-8.

Chuengsamarn, Somlak, Suthee Rattanamongkolgul, Rataya Luechapudiporn, Chada Phisalaphong, and Siwanon Jirawatnotai. 2012. "Curcumin extract for prevention of type 2 diabetes." *Diabetes Care* 35 (11): 2121–2127. doi:10.2337/dc12-0116.

Cooperation, Novartis Pharmaceuticals. 2017. "STARLIX." 3. Accessed 03 10, 2022. https://www.accessdata.fda.gov/drugsatfda_docs/label/2017/021204s015lbl.pdf.

Costa, Geison M, Eloir P Schenkel, and Flávio H Reginatto. 2011. "Chemical and pharmacological aspects of the genus *Cecropia*." *Natural Product Communications* 6 (6): 913–920. doi:10.1177/1934578X1100600637.

Council of Scientific & Industrial Research. 1992. *The Wealth of India: A Dictionary of Indian Raw Materials and Industrial Products, Raw Materials*. Edited by G P Phondke. 3 (Ca-Ci). New Delhi: Publications & Information Directorate, Council of Scientific & Industrial Research.

Crans, Debbie C. 2005. "Fifteen years of dancing with vanadium." *Pure and Applied Chemistry* 77 (9): 1497–1527. doi:10.1351/Pac200577091497.

Crans, Debbie, Barry Posner, LaRee Henry, and Gabriel Cardiff. 2019. "Developing vanadium as an antidiabetic or anticancer drug: A clinical and historical perspective." *Metal Ions in Life Sciences* 19, 203–230.

Cruz, Elda C, and Adolfo Andrade-Cetto. 2015. "Ethnopharmacological field study of the plants used to treat type 2 diabetes among the Cakchiquels in Guatemala." *Journal of Ethnopharmacology* 159: 238–244. doi:10.1016/j.jep.2014.11.021.

Dabhi, Ajay S, Nikita R Bhatt, and Mohit J Shah. 2013. "Voglibose: An alpha glucosidase inhibitor." *Journal of Clinical and Diagnostic Research* 7 (12): 3023–3027. doi:10.7860/JCDR/2013/6373.3838.

Dabrowiak, James C. 2017. *Metals in Medicine*. Hoboken, NJ: John Wiley & Sons Ltd.

Damayanti, Dini S, Didik H Utomo, and Chandra Kusuma. 2017. "Revealing the potency of *Annona muricata* leaves extract as FOXO1 inhibitor for diabetes mellitus treatment through computational study." *In Silico Pharmacology* 5 (1). doi:10.1007/s40203-017-0023-3.

Das, Sayan, Sanjeevani Chaware, Nimish Narkar, Abhijeet V Tilak, Siddhi Raveendran, and Pratik Rane. 2019. "Antidiabetic activity of *Coriandrum sativum* in streptozotocin induced diabetic rats." *International Journal of Basic & Clinical Pharmacology* 8 (5): 925–929. doi:10.18203/2319-2003.ijbcp20191577.

Dash, Gouri K, P Suresh, and S Ganapaty. 2001. "Studies on hypoglycaemic and wound healing activities of *Lantana camara* Linn." *Journal of Natural Remedies* 1 (2): 105–110. doi:10.18311/jnr/2001/16.

Day, C, and C J Bailey. 1986. "Traditional use of *Allium cepa* in the treatment of diabetes mellitus (Abstract)." *Diabetic Med* 3: 361A.

DeFilipps, Robert A, Shirley L Maina, and Juliette Crepin. 2004. *Medicinal Plants of the Guianas (Guyana, Surinam, French Guiana)*. Department of Botany, National Museum of Natural History, Smithsonian Institution. http://botany.si.edu/bdg/medicinal/index.html.

Devi, Sangeeta, and Muneesh Kumar. 2017. "In-vivo antidiabetic activity of methanolic extract of *Euphorbia hirta* L." *International Journal of Diabetes and Endocrinology* 2 (3): 36–39. doi:10.11648/J.IJDE.20170203.11.

Devi, Thiyam R, Davina Hijam, Abhishek Dubey, Suman Debnath, Prabita Oinam, N G T Devi, and W G Singh. 2016. "Study of serum zinc and copper levels in type 2 diabetes mellitus." *International Journal of Contemporary Medical Research* 3 (4): 1036–1040.

Dewi, Rizna T, and Faiza Maryani. 2015. "Antioxidant and α-glucosidase inhibitory compounds of *Centella asiatica*." *Procedia Chemistry* 17: 147–152. doi:10.1016/j.proche.2015.12.130.

Dey, Lucy, Anoja S Attele, and Chun-Su Yuan. 2002. "Alternative therapies for type 2 diabetes." *Alternative Medicine Review* 7 (1): 45–58.

Dhawan, B N, G K Patnaik, R P Rastogi, K K Singh, and J S Tandon. 1977. "Screening of indian plants for biological activity: Part VI." *Indian Journal of Experimental Biology* 15 (3): 208–219.

Dholi, Shravan K, Ramakrishna Raparla, Santhosh K Mankala, and Kannappan Nagappan. 2011. "In vivo antidiabetic evaluation of neem leaf extract in alloxan induced rats." *Journal of Applied Pharmaceutical Science* 1 (4): 100–105.

Diabetes Self Management. 2015. *Diabetes Medicine: Meglitnides*. Chicago, USA: Diabetes Self Management. https://www.diabetesselfmanagement.com/blog/diabetes-medicine-meglitinides/.

Domingo, José L. 2002. "Vanadium and tungsten derivatives as antidiabetic agents: A review of their toxic effects." *Biological Trace Element Research* 88 (2): 97–112. doi:10.1385/BTER:88:2:097.

Douros, Antonios, Hui Yin, Oriana H Y Yu, Kristian B Filion, Laurent Azoulay, and Samy Suissa. 2017. "Pharmacologic differences of sulfonylureas and the risk of adverse cardiovascular and hypoglycemic events." *Diabetes Care* 40 (11): 1506–1513. doi:10.2337/dc17-0595.

Douros, Antonios, Sophie Dell'Aniello, Oriana H Y Yu, Kristian B Filion, L Azoulay, and Samy Suissa. 2018. "Sulfonylureas as second line drugs in type 2 diabetes and the risk of cardiovascular and hypoglycaemic events: Population based cohort study." *British Medical Journal* 362, 1–13. doi:10.1136/bmj.k2693.

Drug Law Center. 2017. *Januvia*. Chicago, USA: Drug Law Center.

DrugLib.com. 2009. *Glucophage (Metformin Hydrochloride) - Side Effects and Adverse Reactions*. 02 27. Accessed 03 14, 2022. http://www.druglib.com/druginfo/glucophage/side-effects_adverse-reactions/.

Eddouks, Mohamed, A Lemhadri, and J B Michel. 2004. "Caraway and caper: Potential anti-hyperglycaemic plants in diabetic rats." *Journal of Ethnopharmacology* 94 (1): 143–148. doi:10.1016/j.jep.2004.05.006.

Egharevba, Ewaen, Patience Chukwuemeke-Nwani, Uche Eboh, Esther Okoye, Israel O Bolanle, Irene O Oseghale, Vincent O Imieje, Osayemwenre Erharuyi, and Abiodun Falodun. 2019. "Evaluation of the antioxidant and hypoglycaemic potentials of the leaf extracts of *Stachytarphyta jamaicensis* (Verbenaceae)." *Tropical Journal of Natural Product Research* 3 (5): 170–174. doi:10.26538/tjnpr/v3i5.4.

Egunyomi, A, Idayat Gbadamosi, and M O Animashahun. 2011. "Hypoglycaemic activity of the ethanol extract of *Ageratum conyzoides* Linn. shoots on alloxan-induced diabetic rats." *Journal of Medicinal Plant Research* 5 (22): 5347–5350. doi:10.5897/JMPR.9000709.

Eidi, Maryam, Akram Eidi, Ali Saeidi, Saadat Molanaei, Alireza Sadeghipour, Massih Bahar, and Kamal Bahar. 2009. "Effect of coriander seed (*Coriandrum sativum* L.) ethanol extract on insulin release from pancreatic beta cells in streptozotocin-induced diabetic rats." *Phytotherapy Research* 23 (3): 404–406. doi:10.1002/ptr.2642.

Ekaiko, Marshall U, Ndulaka J C, Ogbonna C R, and Asiegbu E I. 2016. "Anti-diabetic effect of *Ocimum gratissimum* on blood glucose level in alloxan induced diabetic rats." *IOSR Journal of Pharmacy and Biological Sciences* 11 (2): 12–14. doi:10.9790/3008-1102031214.

El-Demerdash, F M, M I Yousef, N I, and A El-Naga. 2005. "Biochemical study on the hypoglycemic effects of onion and garlic in alloxan-induced diabetic rats." *Food and Chemical Toxicology* 43 (1): 57–63. doi:10.1016/j.fct.2004.08.012.

El-Gamal, and El-Khedr M M. 2017. "Therapeutic benefits of garlic against alloxan-induced diabetic in rats." *Journal of Medical Science And Clinical Research* 5 (2): 17445–17453. doi:10.18535/jmscr/v5i2.39.

El-Moselhy, Mohamed A, Ashraf Taye, Sara S Sharkawi, Suzan F I El-Sisi, and Ahmed F Ahmed. 2011. "The antihyperglycemic effect of curcumin in high fat diet fed rats. Role of TNF-α and free fatty acids." *Food and Chemical Toxicology* 49 (5): 1129–1140. doi:10.1016/j.fct.2011.02.004.

Emran, Talha B, Mycal Dutta, Mir M N Uddin, Aninda K Nath, and Md Z Uddin. 2015. "Antidiabetic potential of the leaf extract of *Centella asiatica* in alloxaninduced diabetic rats." *Journal of Biological Sciences* 4 (1): 51–59. doi:10.3329/jujbs.v4i1.27785.

Estella, Odoh U, Ezugwu C Obodoike, and Udofot E Esua. 2020. "Evaluation of the anti-diabetic and toxicological profile of the leaves of *Stachytarpheta jamaicensis* (L.) Vahl (Verbenaceae) on alloxan-induced diabetic rats." *Journal of Pharmacognosy and Phytochemistry* 9 (3): 477–484.

Evangelopoulos, Angelos, Aris Kollas, Natalia Vallianou, and Christos Kazazis. 2014. "Curcumin and diabetes: Mechanisms of action and its anti-diabetic properties." *Current Topics in Nutraceutical Research* 12 (4): 135–141.

Eyo, J E, and J N Chukwu. 2016. "Antidiabetic potentials of combined aqueous extracts of *Gongronema latifolium* and *Ocimum gratissimum* on alloxan-induced diabetic *Rattus novergicus*." *PharmacologyOnline* 2: 105–115.

Ezenwa, K C, Ighodaro Igbe, and Macdonald Idu. 2015. "Antidiabetic Activity of Methanol extract of *Stachytarpheta jamaicensis* in streptozotocin-induced diabetic rats." *Nigerian Journal of Pharmaceutical Sciences* 14 (1): 9–18.

Fandohan, P, B Gnonlonfin, A Laleye, J Gbenou, R Darboux, and M Moudachirou. 2008. "Toxicity and gastric tolerance of essential oils from *Cymbopogon citratus*, *Ocimum gratissimum* and *Ocimum basilicum* in Wistar rats." *Food and Chemical Toxicology* 46 (7): 2493–2497.

Fatema, Kaniz, Farzana Rahman, Nurunnahar Sumi, Khadizatul Kobura, Afsana Afroz, and Liaquat Ali. 2011. "Glycemic and insulinemic responses to pumpkin and unripe papaya in type 2 diabetic subjects." *International Journal of Nutrition and Metabolism* 3 (1): 1–6. doi:10.5897/IJNAM.9000018.

Fern, Ken. n.d. *Useful Tropical Plants: Citrus aurantiifolia*. Tropical Plants Database. Accessed October 05, 2020. http://tropical.theferns.info/viewtropical.php?id=Citrus+aurantiifolia.

Fillat, Cristina, Joan E Rodriguez-Gil, and J J Guinovart. 1992. "Molybdate and tungstate act like vanadate on glucose metabolism in isolated hepatocytes." *Biochemical Journal* 282 (3): 659–663. doi:10.1042/bj2820659.

Fitrianda, Eka, Elin Y Sukandar, and Elfahmi, I K Adnaya. 2017. "Antidiabetic activity of extract, fractions and asiaticoside compound isolated from *Centella Asiatica* Linn. Leaves in alloxan-induced diabetic mice." *Asian Journal of Pharmaceutical and Clinical Research* 10 (10): 268–272. doi:10.22159/ajpcr.2017.v10i10.20419.

Florence, Ngueguim T, Massa Z Benoit, Kouamouo Jonas, Tchuidjang Alexandra, Dzeufiet D P Désiré, Kamtchouing Pierre, and Dimo Théophile. 2014. "Antidiabetic and antioxidant effects of *Annona muricata* (Annonaceae), aqueous extract on streptozotocin-induced diabetic rats." *Journal of Ethnopharmacology* 151 (2): 784–790. doi:10.1016/j.jep.2013.09.021.

Formigoni, Maria L O S, Helena M Lodder, Oswaldo G Filho, Tania M S Ferreira, and E A Carlini. 1986. "Pharmacology of lemongrass (*Cymbopogan citratus* Staph). II. Effects of daily two month andministration in male and female rats and in offspring exposed 'In Utero'." *Journal of Ethnopharmacology* 17 (1): 65–74. doi:10.1016/0378-8741(86)90073-5.

Galan, Nicole. 2019. "What happens when a person takes too much zinc?" *MedicalNewsToday.* October 23. https://www.medicalnewstoday.com/articles/326760.

Gallwitz, Baptist. 2007. "Review of sitagliptin phosphate: A novel treatment for type 2 diabetes." *Vascular Health and Risk Management* 3 (2): 203–210. doi:10.2147/vhrm.2007.3.2.203.

Ganesh, Thangavel, Saikat Sen, E Thilagam, G Thamotharan, T Loganathan, and Raja Chakraborty. 2010. "Pharmacognostic and anti-hyperglycemic evaluation of *Lantana camara* (L.) var. aculeate leaves in alloxan-induced hyperglycemic rats." *International Journal of Research in Pharmaceutical Sciences* 1 (3): 247–252.

Garba, Husaina A, Aminu Mohammed, Mohammed A Ibrahim, and Mohammed N Shuaibu. 2020. "Effect of lemongrass (*Cymbopogon citratus* Stapf) tea in a type 2 diabetes rat model." *Clinical Phytoscience* 6 (19): 1–10.

Gauri, Munawwar, S J Ali, and Mohd S Khan. 2015. "A review of *Apium graveolens* (Karafs) with special reference to unani medicine." *International Archives of Integrated Medicine* 2 (1): 131–136.

Ghannam, Nadia, Michael Kingston, Ibrahim A Al-Meshaal, Mohamed Tariq, Narayan S Parman, and Nicholas Woodhouse. 1986. "The antidiabetic activity of aloes: Preliminary clinical and experimental observations." *Hormone Research* 24 (4): 288–294. doi:10.1159/000180569.

Ghosh, Abhijeet, and Abhijit Dutta. 2017. "Antidiabetic effects of ethanolic flower extract of *Hibiscus rosa sinensis* (L) on alloxan induced diabetes in hyperlipidaemic experimental Wister rats (WNIN)." *International Journal of Engineering Development and Research* 5 (4): 674–679.

GlaxoSmithKline. 2007. "AVANDIA." Accessed 10 03, 2022. https://www.accessdata.fda.gov/drugsatfda_docs/label/2007/021071s031lbl.pdf.

Goel, Ajay, Ajaikumar B Kunnumakkara, and Bharat B Aggarwal. 2008. "Curcumin as "Curecumin": From kitchen to clinic." *Biochemical Pharmacology* 75 (4): 787–809. doi:10.1016/j.bcp.2007.08.016.

Gomathy, R, N R Vijayalekshmi, and P A Kurup. 1990. "Hypoglycemic action of the pectin present in the juice of the inflorescence stalk of plantain (*Musa sapientum*)—Mechanism of action." *Journal of Biosciences* 15 (4): 297–303.

Graham, David J, Lanh Green, John R Senior, and Parivash Nourjah. 2003. "Troglitazone-induced liver failure: A case study." *The American Journal of Medicine* 114 (4): 299–306. doi:10.1016/s0002-9343(02)01529-2.

Gray, Alison M, and Peter R Flatt. 1999. "Insulin-releasing and insulin-like activity of the traditional anti-diabetic plant *Coriandrum sativum* (coriander)." *British Journal of Nutrition* 81 (3): 203–209. doi:10.1017/s0007114599000392.

Grover, Jagdish K, Satyapal S Yadav, and Vikrant J Vats. 2002. "Hypoglycemic and anti-hyperglycemic effect of *Brassica juncea* diet and their effect on hepatic glycogen content and the key enzymes of carbohydrate metabolism." *Molecular and Cellular Biochemistry* 241 (1–2): 95–101. doi:10.1023/a:1020814709118.

Guardado-Mendoza, Rodolfo, Annamaria Prioletta, Lilia M Jiménez-Ceja, Aravind Sosale, and Franco Folli. 2013. "The role of nateglinide and repaglinide, derivatives of meglitinide, in the treatment of type 2 diabetes mellitus." *Archives of Medical Science* 9 (5): 936–943. doi:10.5114/aoms.2013.34991.

Guinovart Cirera, Joan J, Albert Barbera Lluis, and Joan E Rodriguez Gil. 1997. *Tungsten (VI) Compositions for the Oral treatment of Diabetes Mellitus.* Spain Patent EP 0 755 681 A1. 01 29.

Gundling, Katherine E. 1998. "When did I become an 'allopath'?" *Archives of Internal Medicine* 158 (20): 2185–2186. doi:10.1001/archinte.158.20.2185.

Gungurthy, Jayasree, Sunanda Sabbithia, Krishna Chaitanya, Alekhya Ravella, and C Ramesh. 2013. "Antidiabetic activity of *Leonotis neptefolia* Linn in alloxan induced diabetic rats." *International Journal of Preclinical & Pharmaceutical Research* 4 (1): 5–9.

Gupta, Mahabir P, Nilka G Solis, Mario E Avella, and Ceferino Sanchez. 1984. "Hypoglycemic activity of *Neurolaena lobata* (L.) R. BR." *Journal of Ethnopharmacology* 10 (3): 323–327. doi:10.1016/0378-8741(84)90020-5.

Gutierrez, Rosa M P, Vicente A Juarez, Jahel V Sauceda, and Irasema Anaya-Sosa. 2014. "In vitro and in vivo antidiabetic and antiglycation properties of *Apium graveolens* in type 1 and 2 diabetic rats." *International Journal of Pharmacology* 10 (7): 368–379. doi:10.3923/ijp.2014.368.379.

Hamzah, Rabiat U, Adebimpe A Odetola, Ochuko L Erukainure, and Ademola A Oyagbemi. 2012. "*Peperomia pellucida* in diets modulates hyperglyceamia, oxidative stressand dyslipidemia in diabetic rats." *Journal of Acute Disease* 2012: 135–140. doi:10.1016/S2221-6189(13)60074-8.

Hayanga, Julia A, Senelisiwe P Ngubane, Alfred N Murunga, and Peter M O Owira. 2015. "Grapefruit juice improves glucose intolerance in streptozotocininduced diabetes by suppressing hepatic gluconeogenesis." *European Journal of Nutrition* 55 (2), 631–638. doi:10.1007/s00394-015-0883-4.

Hedayati, Narges, Mehri B Naeini, Arash Mohammadinejad, and Seyed A Mohajeri. 2019. "Beneficial effects of celery (*Apium graveolens*) on metabolic syndrome: A review of the existing evidences." *Phytotherapy Research* 33 (12): 1–14. doi:10.1002/ptr.6492.

Heidari, Himan, Mohammad Kamalinejad, Maryam Noubarani, Mokhtar Rahmati, Iman Jafarian, Hasan Adiban, and Mohammad R Eskandari. 2016. "Protective mechanisms of *Cucumis sativus* in diabetes-related models of oxidative stress and carbonyl stress." *BioImpacts* 6 (1): 33–39. doi:10.15171/bi.2016.05.

Helmstädter, A. 2008. "*Syzygium cumini* (L.) SKEELS (Myrtaceae) against diabetes –125 years of research." *Pharmazie* 63 (2): 91–101.

Herowati, Rina, Levi Puradewa, Juvita Herdianty, and Gunawan P Widodo. 2020. "Antidiabetic activity of okra fruit (*Abelmoschus esculentus* (L) Moench) extract and fractions in two conditions of diabetic rats." *Indonesian Journal of Pharmacy* 31 (1): 27–34.

Horsfall, A U, Olaleye Olabiyi, Ayoola Aiyegbusi, C C Noronha, and Abayomi O Okanlawon. 2008. "*Morinda citrifolia* fruit juice augments insulin action in Sprague-Dawley rats with experimentally induced diabetes." *Nigerian Quarterly Journal of Hospital Medicine* 18 (3): 162–165. doi:10.4314/nqjhm.v18i3.45020.

Hsu, Yi-Jou, Tsung-Han Lee, Cicero L T Chang, Yuh-Ting Huang, and Wen-Chin Yang. 2009. "Anti-hyperglycemic effects and mechanism of *Bidens pilosa* water extract." *Journal of Ethnopharmacology* 122 (2): 379–383. doi:10.1016/j.jep.2008.12.027.

Huseini, Hassan F, Shirin Hasani-Rnjbar, Neda Nayebi, Ramin Heshmat, Farahnaz K Sigaroodi, Maryam Ahvazi, Behroz A Alaei, and Saeed Kianbakht. 2013. "*Capparis spinosa* L. (Caper) fruit extract in treatment of type 2 diabetic patients: A randomized double-blind placebo-controlled clinical trial." *Complementary Therapies in Medicine* 21 (5): 447–452. doi:10.1016/j.ctim.2013.07.003.

Husna, Fauzul, Franciscus D Suyatna, Wawaimuli Arozal, and Erni H Poerwaningsih. 2018. "Anti-diabetic potential of *Murraya Koenigii* (L.) and its antioxidant capacity in nicotinamide-streptozotocin induced diabetic rats." *Drug Research (Stuttgart)* 68 (11): 631–636. doi:10.1055/a-0620-8210.

Hussain, Halim E M A. 2002. "Hypoglycemic, hypolipidemic and antioxidant properties of combination of curcumin from *Curcuma longa*, Linn, and partially purified product from *Abroma augusta*, Linn. in streptozotocin induced diabetes." *Indian Journal of Clinical Biochemistry* 17 (2): 33–43. doi:10.1007/BF02867969.

Ibrahim, Fatima A, Lamidi A Usman, Jubril O Akolade, Oluwafemi A Idowu, Azeemat T Abdulazeez, and Aliyu O Amuzat. 2019. "Antidiabetic potentials of *Citrus aurantifolia* leaf essential oil." *Drug Research* 69 (4): 201–206. doi:10.1055/a-0662-5607.

2021. *IDF Diabetes Atlas 2021: 10th Edition*. International Diabetes Federation. https://diabetesatlas.org/resources/.

Ijioma, Solomon N, A I Okafor, P I Ndukuba, A A Nwankwo, and S C Akomas. 2014. "Hypoglycemic, hematologic and lipid profile effects of *Chromolaena odorata* ethanol leaf extract in alloxan induced diabetic rats." *Annals of Biological Sciences* 2 (3): 27–32.

Islam, Muhammad T, and Mohammad A Uddin. 2017. "A revision on *Urena lobata* L." *International Journal of Medicine* 5 (1): 126–131. doi:10.14419/ijm.v5i1.7525.

Ismawanti, Zuhria, Joseph B Suparyatmo, and Budiyanti Wiboworini. 2019. "The effects of papaya fruit as anti diabetes type 2: A review." *International Journal of Nutrition Sciences* 4 (2): 65–70. doi:10.30476/IJNS.2019.81751.1013.

Jagetia, Ganesh C. 2018. "A review on the role of jamun, *Syzygium cumini* skeels in the treatment of diabetes." *International Journal of Complementary & Alternative Medicine* 11 (2): 91–95. doi:10.15406/IJCAM.2018.11.00374.

Jain, R C, and C R Vyas. 1974. "Hypoglycaemic action of onion on rabbits." *British Medical Journal* 2 (5921): 730. doi:10.1136/bmj.2.5921.730-b.

Jain, R C, and C R Vyas. 1975. "Garlic in alloxan-induced diabetic rabbits." *American Journal of Clinical Nutrition* 28 (7): 684–685. doi:10.1093/ajcn/28.7.684.

Jain, R C, C R Vyas, and O P Mahatma. 1973. "Hypoglycaemic action of onion and garlic." *The Lancet* 2 (7844): 1491. doi:10.1016/s0140-6736(73)92749-9.

Jain, Sushil K, Justin Rains, Jennifer Croad, Bryon Larson, and Kimberly Jones. 2009. "Curcumin supplementation lowers TNF-α, IL-6, IL-8, and MCP- 1 secretion in high glucose-treated cultured monocytes and blood levels of TNF-α, IL-6, MCP-1, glucose, and glycosylated hemoglobin in diabetic rats." *Antioxidants & Redox Signaling* 11 (2): 241–249. doi:10.1089/ars.2008.2140.

Jaiswal, Dolly, Prashant K Rai, Amit Kumar, Shikha Mehta, and Geeta Watal. 2009. "Effect of *Moringa oleifera* Lam. leaves aqueous extract therapy on hyperglycemic rats." *Journal of Ethnopharmacology* 123 (3): 392–396. doi:10.1016/j.jep.2009.03.036.

Jaleel, C A, R Gopi, G M A Lakshmanan, and R Panneerselvam. 2006. "Triadimefon induced changes in the antioxidant metabolism and ajmalicine production in *Catharanthus roseus* (L.) G. Don." *Plant Science* 171 (2): 271–276.

Jansen, Judith, Wolfram Karges, and Lothar Rink. 2009. "Zinc and diabetes - clinical links and molecular mechanisms." *Journal of Nutritional Biochemistry* 20 (6): 399–417. doi:10.1016/j.jnutbio.2009.01.009.

Janssen Pharmaceuticals, Inc. 2018. "INVOKANA." 10. Accessed 03 10, 2022. https://www.accessdata.fda.gov/drugsatfda_docs/label/2018/204042s027lbl.pdf.

Jawonisi, Ibitade O, and Godwin I Adoga. 2015. "Hypoglycaemic and hypolipidaemic effect of extract of *Lantana camara* Linn. leaf on alloxan diabetic rats." *Journal of Natural Sciences Research* 5 (8): 57–65.

Jenei, Kinga, Ildikó Szatmári, Eszter Szabó, Anjum Mariam, Andrea Luczay, Petra Zsidegh, and Péter Tóth-Heyn. 2019. "Changes of lactate levels in diabetic ketoacidosis and in newly diagnosed type 1 diabetes mellitus." *Orvosi Hetilap* 160 (45): 1784–1790.

Jordan, M A, D Thrower, and L Wilson. 1991. "Mechanism of inhibition of cell proliferation by Vinca alkaloids." *Cancer Research* 51 (8): 2212–2222.

Joseph, B, and S J Raj. 2011. "An overview: Pharmacognostic properties of *Phyllanthus amarus* Linn." *International Journal of Pharmacology* 7 (1): 40–45. doi:10.3923/ijp.2011.40.45.

Joseph, Baby, and D Jini. 2013. "Antidiabetic effects of *Momordica charantia* (bitter melon) and its medicinal potency." *Asian Pacific Journal of Tropical Disease* 3 (2): 93–102. doi:10.1016/S2222-1808(13)60052-3.

Joshi, Archana, R K Bachheti, Ashutosh Sharma, and Ritu Mamgain. 2016. "*Parthenium hysterophorus*. L. (Asteraceae): A boon or curse? (A review)." *Oriental Journal of Chemistry* 32 (3): 1283–1294. doi:10.13005/ojc/320302.

Jovanovic, Lois, Mario Gutierrez, and Charles M Peterson. 1999. "Chromium supplementation for women with gestational diabetes mellitus." *The Journal of Trace Elements in Experimental Medicine* 12 (2): 91–97. doi:10.1002/(SICI)1520-670X(1999)12:2<91::AID-JTRA6>3.0.CO;2-X.

Juárez-Rojop, Isela E, Juan C Díaz-Zagoya, Jorge L Ble-Castillo, Pedro H Miranda-Osorio, Andrés E Castell-Rodríguez, Carlos A Tovilla-Zárate, and Deysi Y Bermúdez-Ocaña. 2012. "Hypoglycemic effect of *Carica papaya* leaves in streptozotocin-induced diabetic rats." *BMC Complementary and Alternative Medicine* 12 (1): 1–11. doi:10.1186/1472-6882-12-236.

Jung, Un J, Mi-Kyung Lee, Kyu-Shik Jeong, and Myung-Sook Choi. 2004. "The hypoglycemic effects of hesperidin and naringin are partly mediated by hepatic glucose-regulating enzymes in C57BL/KsJ-db/db mice." *The Journal of Nutrition* 134 (10): 2499–2503. doi:10.1093/jn/134.10.2499.

Kahn, C R, Lihong Chen, and Shmuel E Cohen. 2000. "Unraveling the mechanism of action of thiazolidinediones." *The Journal of Clinical Investigation* 106 (11): 1305–1307. doi:10.1172/JCI11705.

Kalita, Sanjeeb, Gaurav Kumar, Loganathan Karthik, and Kokati V B Rao. 2012. "A review on medicinal properties of *Lantana camara* Linn." *Research Journal of Pharmacy and Technology* 5 (6): 711–715.

Kalra, Sanjay, S V Madhu, and Sarita Bajaj. 2015. "Sulfonylureas: Assets in the past, present and future." *Indian Journal of Endocrinology and Metabolism* 19 (3): 314–316. doi:10.4103/2230-8210.149925.

Kanedi, Mohammad, Sutyarso Hendri Busman, Cahyani I Kesuma, Yulianty, and Martha L Lande. 2019. "Ameliorative effect of plant extracts of suruhan (*Peperomia pellucida*) on blood glucose and libido of male mice injected with alloxan." *European Journal of Biomedical and Pharmaceutical Sciences* 6 (2): 18–21.

Kanedi, Mohammad, Sutyarso Hendri Busman, Risky, A Mandasari, and Gina D Pratami. 2019. "Plant extracts of Suruhan (*Peperomia pellucida* L. Kunth) ameliorate infertility of male mice with alloxan-induced hyperglycemia." *International Journal of Biomedical Research* 10 (2): 1–4. doi:10.7439/ijbr.

Karimi, Akbar, and Nahid K Nasab. 2014. "Effect of garlic extract and *Citrus aurantifolia* (lime) juice and on blood glucose level and activities of aminotransferase enzymes in streptozotocin-induced diabetic rats." *World Journal of Pharmaceutical Sciences* 2 (8): 821–827.

Karthiyayini, T, Rajesh Kumar, K L S Kumar, Ram K Sahu, and Amit Roy. 2009. "Evaluation of antidiabetic and hypolipidemic effect of *Cucumis sativus* fruit in streptozotocin-induced-diabetic rats." *Biomedical and Pharmacology Journal* 2 (2): 351–355.

Kawamori, Ryuzo, Kohei Kaku, Toshiaki Hanafusa, Daisuke Kashiwabara, Shigeru Kageyama, and Nigishi Hotta. 2012. "Efficacy and safety of repaglinide vs nateglinide for treatment of Japanese patients with type 2 diabetes mellitus." *Journal of Diabetes Investigation* 3 (3): 302–308. doi:10.1111/j.2040-1124.2011.00188.x.

Kawamori, Ryuzo, Naoko Tajima, Yasuhiko Iwamoto, Atsunori Kashiwagi, Kazuaki Shimamoto, and Kohei Kaku. 2009. "Alpha-glucosidase inhibitor for the prevention of type 2 diabetes mellitus: A randomised, double-blind trial in japanese subjects with impaired glucose tolerance." *Nihon rinsho. Japanese Journal of Clinical Medicine* 67 (9): 1821–1825. https://www.ncbi.nlm.nih.gov/pubmed/19768922.

Kazmi, Imran, Mahfoozur Rahman, Muhammad Afzal, Gaurav Gupta, Shakir Saleem, Obaid Afzal, Adil Shaharyar, Ujjwal Nautiyal, Sayeed Ahmed, and Firoz Anwar. 2012. "Anti-diabetic potential of ursolic acid stearoyl glucoside: A new triterpenic gycosidic ester from *Lantana camara*." *Fitoterapia* 83 (1): 142–146. doi:10.1016/j.fitote.2011.10.004.

Kesari, Achyut N, Shweta Kesari, Santosh K Singh, Rajesh K Gupta, and Geeta Watal. 2007. "Studies on the glycemic and lipidemic effect of *Murraya koenigii* in experimental animals." *Journal of Ethnopharmacology* 112 (2): 305–311. doi:10.1016/j.jep.2007.03.023.

Khaket, Tejinder P, Himanshu Aggarwal, Drukshakshi Jodha, Suman Dhanda, and Jasbir Singh. 2015. "*Parthenium hysterophorus* in current scenario: A toxic weed with industrial, agricultural and medicinal applications." *Journal of Plant Sciences* 10 (2): 42–53. doi:10.3923/jps.2015.42.53.

Khan, B A, A Abraham, and S Leelamma. 1997. "Anti-oxidant effects of curry leaf, *Murraya koenigii* and mustard seeds, *Brassica juncea* in rats fed with high fat diet." *Indian Journal of Experimental Biology* 35 (2): 148–150.

Khan, Mohd. Y, Irfan Aziz, Bipin Bihari, Hemant Kumar, Maryada Roy, and Vikas K Verma. 2014. "A review - Phytomedicines used in treatment of diabetes." *International Journal of Pharmacognosy* 1 (6): 343–365. doi:10.13040/IJPSR.0975-8232.1(6).343-65.

Khatoon, Rahmat T, Surendar Angothu, Y Krishnaveni, and C M Reddy. 2013. "Investigation of antidiabetic activity of *Lantana camara* Linn. (Verbenaceae) leaves." *Internation Journal of Phytopharmacy Research* 4 (1): 1–4.

Khatun, Hajera, Ajijur Rahman, Mohitosh Biswas, and Anwar Ul Islam. 2011. "Water-soluble fraction of *Abelmoschus esculentus* L interacts with glucose and metformin hydrochloride and alters their absorption kinetics after coadministration in rats." *ISRN Pharmaceutics* 2011: 1–5. doi:10.5402/2011/260537.

Kim, Jong-Hak, Min-Kyeoung Kim, Hongtao Wang, Hee-Nyeong Lee, Chi-Gyu Jin, Woo-Saeeng Kwon, and Deok-Chun Yang. 2016. "Discrimination of Korean ginseng (*Panax ginseng* Meyer) cultivar Chunpoong and American ginseng (*Panax quinquefolius*) using the auxin repressed protein gene." *Journal of Ginseng Research* 40 (4): 395–399. doi:10.1016/j.jgr.2015.12.002.

Kim, Kang Hoon, In-Seung Lee, Ji Young Park, Yumi Kim, Eun-Jin An, and Hyeung-Jin Jang. 2018. "Cucurbitacin B induces hypoglycemic effect in diabetic mice by regulation of AMP-activated protein kinase alpha and glucagon-like peptide-1 via bitter taste receptor signaling." *Frontiers in Pharmacology*. doi:10.3389/fphar.2018.01071.

Kingma, P J, P P Menheere, J P Sels, and A C N Kruseman. 1992. "Alpha-glucosidase inhibition by miglitol in NIDDM patients." *Diabetes Care* 15 (4): 478–483. doi:10.2337/diacare.15.4.478.

Koffi, N'guessan, Kouassi K Édouard, and Kouadio Kouassi. 2009. "Ethnobotanical study of plants used to treat diabetes, in traditional medicine, by Abbey and Krobou people of Agboville (Côte-d'Ivoire)." *American Journal of Scientific Research* 4: 45–48.

Kohli, K R, and R H Singh. 1993. "A clinical trial of jambu (*Eugenia jambolana*) in non insulin dependent diabetes mellitus." *Journal of Research in Ayurveda and Siddha* 13: 89–97.

Konda, Prabhakar Y, Janardhan Y Egi, Sreenivasulu Dasari, Raju Katepogu, Krishna K Jaiswal, and Prabhusaran Nagarajan. 2020. "Ameliorative effects of *Mentha aquatica* on diabetic and nephroprotective potential activities in STZ-induced renal injury." *Comparative Clinical Pathology* 29 (2): 189–199. doi:10.1007/s00580-019-03042-6.

Kooti, Wesam, Sara Ali-Akbari, Majid Asadi-Samani, Hosna Ghadery, and Damoon Ashtary-Larky. 2014. "A review on medicinal plant of *Apium graveolens*." *Advanced Herbal Medicine* 1 (1): 48–59.

Kraemer, Fredric B, and Henry N Ginsberg. 2014. "Gerald M. Reaven, MD: Demonstration of the central role of insulin resistance in type 2 diabetes and cardiovascular disease." *Diabetes Care* 37: 1178–1181.

Kulkarni, R N, Kaushik Baskaran, R S Chandrashekara, and S Kumar. 1999. "Inheritance of morphological traits of periwinkle mutants with modified contents and yields of leaf and root." *Plant Breeding* 118 (1): 71–74. doi:10.1046/j.1439-0523.1999.118001071.x.

Kumar, A, R Ilavarasan, T Jayachandran, M Deecaraman, P Aravindan, N Padmanabhan, and M R V Krishan. 2008. "Anti-diabetic activity of *Syzygium cumini* and its isolated compound against streptozotocin-induced diabetic rats." *Journal of Medicinal Plants Research* 2 (9): 246–249.

Kumar, Anil, Rita M Sunday, and Efere M Obuotor. 2019. "Antioxidant and antidiabetic properties of *Mimosa pudica* seeds in streptozotocin-induced diabetic wistar rats." *Asian Journal of Biotechnology* 12 (1): 1–8. doi:10.3923/ajbkr.2020.1.8.

Kumar, Sunil, Rashmi Malhotra, and Dinesh Kumar. 2010a. "Antihyperglycemic, antihyperlipidemic and antioxidant activities of *Euphorbia hirta* stem extract." *International Research Journal of Pharmacy* 1 (1): 150–156.

Kumar, Sunil, Rashmi Malhotra, and Dinesh Kumar. 2010b. "Evaluation of antidiabetic activity of *Euphorbia hirta* Linn. in streptozotocin induced diabetic mice." *Indian Journal of Natural Products and Resources* 1 (2): 200–203.

Kumar, Sunil, Smita Narwal, Vipin Kumar, and Om Prakash. 2011. "α-Glucosidase inhibitors from plants: A natural approach to treat diabetes." *Pharmacognosy Reviews* 5 (9): 19–29. doi:10.4103/0973-7847.79096.

Kumar, Vikas, Ajit K Thakur, Narottam D Barothia, and Shyam S Chatterjee. 2011. "Therapeutic potentials of *Brassica juncea*: An overview." *TANG: International Journal of Genuine Traditional Medicine* 1 (1): 1–17. doi:10.5667/tang.2011.0005.

Kumar, Vishnu, Farzana Mahdi, Ashok K Khanna, Ranjana Singh, Ramesh Chander, Jitendra K Saxena, Abbas A Mahdi, and Raj K Singh. 2013. "Antidyslipidemic and antioxidant activities of *Hibiscus rosa sinensis* root extract in alloxan induced diabetic rats." *Indian Journal of Clinical Biochemistry* 28 (1): 46–50. doi:10.1007/s12291-012-0223-x.

Kumari, K, B C Mathew, and K T Augusti. 1995. "Antidiabetic and hypolipidemic effects of S-methyl cysteine sulfoxide isolated from *Allium cepa* Linn." *Indian Journal of Biochemistry and Biophysics* 32 (1): 49–54.

Kwon, Y-I, E Apostolidis, Y-C Kim, and K Shetty. 2007. "Health benefits of traditional corn, beans, and pumpkin: In vitro studies for hyperglycemia and hypertension management." *Journal of Medicinal Food* 10 (2): 266–275. doi:10.1089/jmf.2006.234.

Lai, Bun-Yueh, Tzung-Yan Chen, Shou-Hsien Huang, Tien-Fen Kuo, Ting-Hsiang Chang, Chih-Kang Chiang, Meng-Ting Yang, and Cicero L-T Chang. 2015. "*Bidens pilosa* formulation improves blood homeostasis and β -cell function in men: A pilot study." *Evidence-Based Complementary and Alternative Medicine* (Hindawi Publishing Corporation) 2015 (1): 1–5. doi:10.1155/2015/832314.

Latha, Muniappan, and Leelavinothan Pari. 2004. "Effect of an aqueous extract of *Scoparia dulcis* on blood glucose, plasma insulin and some polyol pathway enzymes in experimental rat diabetes." *Brazilian Journal of Medical and Biological Research* 37 (4): 577–586. doi:10.1590/s0100-879x2004000400015.

Latha, Muniappan, Leelavinothan Pari, Sandhya Sitasawad, and Ramesh Bhonde. 2004. "Insulin-secretagogue activity and cytoprotective role of the traditional antidiabetic plant *Scoparia dulcis* (sweet broomweed)." *Life Sciences* 75 (16): 2003–2014. doi:10.1016/j.lfs.2004.05.012.

Latif, Amira A E, Badr E S E Bialy, Hamada D Mahboub, and Mabrouk A A Eldaim. 2014. "*Moringa oleifera* leaf extract ameliorates alloxan-induced diabetes in rats by regeneration of β cells and reduction of pyruvate carboxylase expression." *Biochemistry and Cell Biology* 92 (5): 413–419. doi:10.1139/bcb-2014-0081.

Lawal, Tajudeen A, Kailani Salawu, Abdullah A Imam, Alhassan M Wudil, and Adamu J Alhassan. 2019. "Effect of aqueous extract of *Ocimum gratissimum* on glycaemic index of starch meal." *Global Journal of Medicinal Plants Research* 5 (3): 9–15.

Lawson-Evi, Povi, Kwashie Eklu-Gadegbeku, Amegnona Agbonon, Kodjo Aklikokou, Edmond Creppy, and Messanvi Gbeassor. 2011. "Antidiabetic activity of *Phyllanthus amarus* Schum and Thonn (Euphorbiaceae) on alloxan induced diabetes in male wistar rats." *Journal of Applied Sciences* 11 (16): 2968–2973. doi:10.3923/jas.2011.2968.2973.

Le, Bach Thi, Dung Tien Le, Tuan Trong Nguyen, Chau Thanh Quoc Nguyen, Quetin-Leclercq Joëlle, and Buu Hue Thi Bui. 2018. "The protective effect of some extracts and isolated compounds from *Euphorbia hirta* on pancreatic B-cells MIN6." *Vietnam Journal of Science and Technology* 56 (4A): 163–170.

Lebovitz, H E. 1998. "α-Glucosidase inhibitors as agents in the treatment of diabetes." *Diabetes Reviews* 6: 132–145.

Lee, So-Young, Jin-Taek Hwang, So-Lim Park, Sung-Hun Yi, Young-Do Nam, and Seong-Il Lim. 2012. "Antidiabetic effect of *Morinda citrifolia* (Noni) fermented by Cheonggukjang in KK-Ay diabetic mice." *Evidence-Based Complementary and Alternative Medicine*, 2012(3): 1–7. doi:10.1155/2012/163280.

Leroux-Stewart, Josée, Rémi Rabasa-Lhoret, and Jean-Loius Chiasson. 2015. "α -Glucosidase inhibitors." Chap. 45 in *International Textbook of Diabetes Mellitus*, edited by Ralph A DeFronzo, Ele Ferrannini, Paul Zimmet and K G M M Alberti, 673–685. Wiley-Blackwell. doi:10.1002/9781118387658.ch45.

Li, Jinping, Gerard Elberg, Jacqueline Libman, Abraham Shanzer, Dov Gefel, and Yoram Shechter. 1995. "Insulin-like effects of tungstate and molybdate: Mediation through insulin receptor independent pathways." *Endocrine* 3: 631–637.

Liao, Xuemei, Yanfei Wang, and Chi-Wai Wong. 2010. "Troglitazone induces cytotoxicity in part by promoting the degradation of peroxisome proliferator-activated receptor γ co-activator-1α protein." *British Journal of Pharmacology* 161 (4): 771–781. doi:10.1111/j.1476-5381.2010.00900.x.

Liew, Pearl M, and Yoke K Yong. 2016. "*Stachytarpheta jamaicensis* (L.) Vahl: From traditional usage to pharmacological evidence." *Evidence-Based Complementary and Alternative Medicine* 2016: 1–7. doi:10.1155/2016/7842340.

Lim, Soo, Ji W Yoon, Sung H Choi, Bong J Cho, Jun T Kim, Ha S Chang, Ho S Park, et al. 2009. "Effect of ginsam, a vinegar extract from *Panax ginseng*, on body weight and glucose homeostasis in an obese insulin-resistant rat model." *Metabolism: Clinical and Experimental* 58 (1): 8–15. doi:10.1016/j.metabol.2008.07.027.

Lin, Boniface J, Huey-Peir Wu, H S Huang, Jyuhn Huarng, A Sison, Dato K bin Abdul Kadir, Chung-Gu Cho, and Witaya Sridama. 2003. "Efficacy and tolerability of acarbose in Asian patients with type 2 diabetes inadequately controlled with diet and sulfonylureas." *Journal of Diabetes and Its Complications* 17 (4): 179–185. doi:10.1016/s1056-8727(02)00258-1.

Liu, Hui-Kang, Brian D Green, Neville H McClenaghan, Jane T McCluskey, and Peter R Flatt. 2004. "Long-term beneficial effects of vanadate, tungstate and molybdate on insulin secretion and function of cultured beta cells." *Journal of the Pancreas* 28 (4): 364–368. doi:10.1097/00006676-200405000-00002.

Loganathan, T, and Raghvendra S Bhadauria. 2019. "Antidiabetic, antihyperlipidaemic activity of methanolic extract of *Lantana camara* L leaves in streptozotocin-induced diabetic rats." *International Journal of Experimental Pharmacology* 9 (2): 34–42.

Luka, C D, and A Mohammed. 2012. "Evaluation of the anti diabetic property of aqueous extract of *Mangifera indica* leaf on normal and alloxan-induced diabetic rats." *Journal of Natural Product and Plant Resources* 2 (2): 239–243.

Lundahl, J, C G Regardh, B Edgar, and G Johnsson. 1997. "Effects of grapefruit juice ingestion - pharmacokinetics and haemodynamics of intravenously and orally administered felodipine in healthy men." *European Journal of Clinical Pharmacology* 52 (2): 139–145. doi:10.1007/s002280050263.

Lv, Juan, Lanxiu Cao, Min Li, Rui Zhang, Fu Bai, and Pengfei Wei. 2018. "Effect of hydroalcohol extract of lemon (Citrus limon) peel on a rat model of type 2 diabetes." *Tropical Journal of Pharmaceutical Research* 17 (7): 1367–1372. doi:10.4314/tjpr.v17i7.20.

Lybrate. 2021. *Voglibose. 3 MG Tablet*. Lybrate. March 10. https://www.lybrate.com/medicine/voglibose-0-3-mg-tablet.

Ma, Chan-Juan Ma, Ai-Fang Nie, Zhi-Jian Zhang, Zhi-Guo Zhang, Li Du, Xiao-Ying Li, and Guang Ning. 2013. "Genipin stimulates glucose transport in C2C12 myotubes via an IRS-1 and calcium-dependent mechanism." *Journal of Endocrinology*, 216(3): 353–362. doi:10.1530/JOE-11-0473.

Machraoui, M, Z Kthiri, M B Jabeur, and W Hamada. 2018. "Ethnobotanical and phytopharmacological notes on *Cymbopogon citratus* (DC.) Stapf." *Journal of New Sciences, Agriculture and Biotechnology* 55 (5): 3642–3652.

Mahabir, D, and M C Gulliford. 1997. "Use of medicinal plants for diabetes in Trinidad and Tobago." *Revista Panamericana de Salud Pública/Pan American Journal of Public Health* 1 (3): 174–179. doi:10.1590/S1020-49891997000300002.

Mahesh, Thirunavukarasu, Murali M S Balasubashini, and Venugopal P Menon. 2004. "Photo-irradiated curcumin supplementation in streptozotocin-induced diabetic rats: Effect on lipid peroxidation." *Thérapie* 59 (6): 639–644. doi:10.2515/therapie:2004110.

Maiti, Rajkumar, D Jana, U K Das, and Debidas Ghosh. 2004. "Antidiabetic effect of aqueous extract of seed of *Tamarindus indica* in streptozotocin-induced diabetic rats." *Journal of Ethnopharmacology* 92 (1): 85–91. doi:10.1016/j.jep.2004.02.002.

Makni, Mohamed, Hamadi Fetoui, Nabil K Gargouri, El M Garoui, and Najiba Zeghal. 2011. "Antidiabetic effect of flax and pumpkin seed mixture powder: Effect on hyperlipidemia and antioxidant status in alloxan diabetic rats." *Journal of Diabetes and its Complications* 25 (5): 339–345. doi:10.1016/j.jdiacomp.2010.09.001.

Malematja, R O, V P Bagla, I Njanje, V Mbazima, K W Poopedi, L Mampuru, and M P Mokgotho. 2018. "Potential hypoglycaemic and antiobesity effects of *Senna italica* leaf acetone extract." *Evidence-Based Complementary and Alternative Medicine* 2018: 1–10. doi:10.1155/2018/5101656.

Malvi, Reetesh, Sonam Jain, Shreya Khatri, Arti Patel, and Smita Mishra. 2011. "A review on antidiabetic medicinal plants and marketed herbal formulations." *International Journal of Pharmaceutical & Biological Archives* 2 (5): 1344–1355.

Mamun, Al, Saiful Islam, A H M K Alam, Aziz A Rahman, and Mamunur Rashid. 2013. "Effects of ethanolic extract of *Hibiscus rosa-sinensis* leaves on alloxan-induced diabetes with dyslipidemia in rats." *Bangladesh Pharmaceutical Journal* 16 (1): 27–31. doi:10.3329/bpj.v16i1.14487.

Maniyar, Yasmeen, and Prabhu Bhixavatimath. 2012. "Antihyperglycemic and hypolipidemic activities of aqueous extract of *Carica papaya* Linn. leaves in alloxan-induced diabetic rats." *Journal of Ayurveda and Integrative Medicine* 3 (2): 70–74. doi:10.4103/0975-9476.96519.

Mans, Kamal, and Talal Aburjai. 2019. "Accessing the hypoglycemic effects of seed extract from celery (*Apium graveolens*) in alloxan-induced diabetic rats." *Journal of Pharmaceutical Research International* 26 (6): 1–10. doi:10.9734/jpri/2019/v26i630152.

Mascolo, Nicola, R Jain, S C Jain, and F Capasso. 1989. "Ethnopharmacologic investigation of ginger (*Zingiber officinale*)." *Journal of Ethnopharmacology* 27 (1–2): 129–140. doi:10.1016/0378-8741(89)90085-8.

Masic, Izet, Armin Skrbo, Nabil Naser, Salih Tandir, Lejla Zunic, Senad Medjedovic, and Aziz Sukalo. 2017. "Contribution of Arabic medicine and pharmacy to the development of health care protection in Bosnia and Herzegovina - The first part." *Medical Archives* 71 (5): 364–372.

Mathew, P T, and K T Augusti. 1973. "Studies on the effect of allicin (diallyl disulphide-oxide) on alloxan diabetes. I. Hypoglycaemic action and enhancement of serum insulin effect and glycogen synthesis." *Indian Journal of Biochemistry and Biophysics* 10 (3): 209–212.

Mathew, P T, and K T Augusti. 1975. "Hypoglycaemic effects of onion, *Allium cepa* Linn. on diabetes mellitus - A preliminary report." *Indian Journal of Physiology and Pharmacology* 19 (4): 213–217.

Maurya, Anup K, Smriti Tripathi, Zabeer Ahmed, and Ram K Sahu. 2012. "Antidiabetic and antihyperlipidemic effect of *Euphorbia hirta* in streptozotocin induced diabetic rats." *Der Pharmacia Lettre* 4 (2): 703–707.

Mawarti, Herin, Mohammad Z A Khotimah, and Mukhammad Rajin. 2018. "Ameliorative effect of *Citrus aurantifolia* and *Cinnamomum burmannii* extracts on diabetic complications in a hyperglycemic rat model." *Tropical Journal of Pharmaceutical Research* 17 (5): 823–829. doi:10.4314/tjpr.v17i5.11.

McFadyen, Rachel C. 1992. "Biological control against parthenium weed in Australia." *Crop Protection* 11 (5): 400–407. doi:10.1016/0261-2194(92)90021-V.

Mea, A, Y H R Ekissi, K J C Abo, and Bi G P Kahou. 2017. "Hypoglycaemiant and anti-hyperglycaemiant effect of Juscticia secunda m. vahl (acanthaceae) on glycaemia in the wistar rat." *International Journal of Development Research* 7 (6): 13178–13184.

Meena, Jyoti, R A Sharma, and Rashmi Rolania. 2018. "A review on phytochemical and pharmacological properties of *Phyllanthus amarus* Schum. and Thonn." *International Journal of Pharmaceutical Sciences and Research* 9 (4): 1377–1386.

Meier, Juris J. 2012. "GLP-1 receptor agonists for individualized treatment of type 2 diabetes mellitus." *Nature Reviews Endocrinology* 8 (12): 728–742. doi:10.1038/nrendo.2012.140.

Mendes, John. 1986. *Cote ce Cote la: Trinidad & Tobago Dictionary.* Arima: New Millennium.

Merk and Company Inc. 2021a. "STEGLATRO." 09. Accessed 03 10, 2022. https://www.merck.com/product/usa/pi_circulars/s/steglatro/steglatro_pi.pdf.

Merk and Company Incorporated. 2021b. "JANUVIA." 12. Accessed 10 03, 2022. https://www.merck.com/product/usa/pi_circulars/j/januvia/januvia_pi.pdf.

Mew, Daphne, Felipe Balza, G H N Towers, and Julia G Levy. 1982. "Anti-tumour effects of the sesquiterpene lactone parthenin." *Planta Medica* 45 (5): 23–27. doi:10.1055/s-2007-971234.

Mishra, Chetna, Monowar Alam Khalid, Nazmin Fatima, Babita Singh, Dinesh Tripathi, Mohammad Waseem, and Abbas Ali Mahdi. 2019. "Effects of citral on oxidative stress and hepatic key enzymes of glucose metabolism in streptozotocin/high-fat-diet induced diabetic dyslipidemic rats." *Iranian Journal of Basic Medical Sciences* 22 (1): 49–57.

Mishra, Manas R, A Mishra, D K Pradhan, Ashis K Panda, R K Behera, and Sh Jha. 2013. "Antidiabetic and antioxidant activity of *Scoparia dulcis* Linn." *Indian Journal of Pharmaceutical Sciences* 75 (5): 610–614. doi:10.4103/0250-474X.122887.

Mishra, Manas R, R K Behera, Sh Jha, Ashis K Panda, A Mishra, D K Pradhan, and P R Choudary. 2011. "A brief review on phytoconstituents and ethnopharmacology of *Scoparia dulcis* Linn. (Scrophulariaceae)." *International Journal of Phytomedicine* 3 (4): 422–438.

Mohammadi, Khosro, Katherine H Thompson, Brian O Patrick, Tim Storr, Candice Martins, Elena Polishchuk, Violet G Yuen, John H McNeill, and Chris Orvig. 2005. "Synthesis and characterization of dual function vanadyl, gallium and indium curcumin complexes for medicinal applications." *Journal of Inorganic Biochemistry* 99 (11): 2217–2225. doi:10.1016/j.jinorgbio.2005.08.001.

Moqbel, Fahmi S, Prakash R Naik, Habeeb M Najma, and S Selvaraj. 2011. "Antidiabetic properties of *Hibiscus rosa sinensis* L. leaf extract fractions on non-obese diabetic (NOD) mouse." *Indian Journal of Experimental Biology* 49 (1): 24–29.

Morales, Cristina, Maria P Gómez-Serranillos, Irene Iglesias, Angel M Villar, and Armando Caceres. 2001. "Preliminary screening of five ethnomedicinal plants of Guatemala." *Il Farmaco* 56 (5–7): 523–526. doi:10.1016/s0014-827x(01)01107-7.

Morgan & Morgan. 2015. *Januvia Lawsuits*. USA: Morgan & Morgan.

Moshi, Mainen J, Janet J K Lutale, Gerald H Rimoy, Zulfikar G Abbas, Robert M Josiah, and Andrew B M Swai. 2001. "The effect of *Phyllanthus amarus* aqueous extract on blood glucose in non-insulin dependent diabetic patients." *Phytotherapy Research* 15 (7): 577–580. doi:10.1002/ptr.780.

Mousavi, Leila, Rabeta M Salleh, and Vikneswaran Murugaiyah. 2018a. "Phytochemical and bioactive compounds identification of *Ocimum tenuiflorum* leaves of methanol extract and its fraction with an anti-diabetic potential." *International Journal of Food Properties* 21 (1): 2390–2399. doi:10.1080/10942912.2018.1508161.

Mousavi, Leila, Rabeta M Salleh, and Vikneswaran Murugaiyah. 2018b. "Toxicology assessment of *Ocimum tenuiflorum* L. leaves extracts on streptozotocin-induced diabetic rats." *Malaysian Journal of Microscopy* 14 (1): 124–156.

Mozersky, R P. 1999. "Herbal products and supplemental nutrients used in the management of diabetes." *The Journal of the American Osteopathic Association* 99 (12) Supplement: S4–S9. doi:10.7556/jaoa.1999.99.12.S4.

Muhlishoh, Arwin, Brian Wasita, and Adi M P Nuhriawangsa. 2018. "Antidiabetic effect of *Centella asiatica* extract (whole plant) in streptozotocin nicotinamide-induced diabetic rats." *Jurnal Gizi dan Dietetik Indonesia (Indonesian Journal of Nutrition and Dietetics)* 6 (1): 14–22. doi:10.21927/ijnd.2018.6(1).14-22.

Murray, Robert K, Victor W Rodwell, David A Bender, Kathleen M Botham, P A Weil, and Peter J Kennelly. 2009. *Harper's Illustrated Biochemistry*. 28th. New York, NY: McGraw-Hill Medical.

Murti, Krishna, Mayank Panchal, Poonam Taya, and Raghuveer Singh. 2012. "Pharmacological properties of *Scoparia dulcis*: A review." *Pharmacologia* 3 (8): 344–347. doi:10.5567/pharmacologia.2012.344.347.

Murugan, Pidaran, and Leelavinothan Pari. 2007. "Influence of tetrahydrocurcumin on hepatic and renal functional markers and protein levels in experimental type 2 diabetic rats." *Basic & Clinical Pharmacology & Toxicology* 101 (4): 241–245. doi:10.1111/j.1742-7843.2007.00109.x.

N, Sujnan, Satish S, Karunakar Hegde, and AR Shabaraya. 2018. "A review on pharmacological activities of the plant *Peperomia pellucida* L." *International Journal of Pharma and Chemical Research* 4 (2): 106–110.

Na, L-X, Y-L Zhang, Y Li, L-Y Liu, R Li, T Kong, and C-H Su. 2011. "Curcumin improves insulin resistance in skeletal muscle of rats." *Nutrition, Metabolism & Cardiovascular Diseases* 21 (7): 526–533. doi:10.1016/j.numecd.2009.11.009.

Nabavi, Seyed F, Filippo Maggi, Maria Daglia, Solomon Habtemariam, Luca Rastrelli, and Seyed M Nabavi. 2016. "Pharmacological effects of *Capparis spinosa* L." *Phytotherapy Research* 30 (11): 1733–1744. doi:10.1002/ptr.5684.

Nadro, M S, and I O Onoagbe. 2014. "Chemical constituents of ethanolic extract of *Cassia italica* leaf." *Asian Journal of Biochemical and Pharmaceutical Research* 4 (3): 328–334.

Naim, Mohammad, Farhad M Amjad, Sania Sultana, Sheikh N Islam, Muhammad A Hossain, Rehana Begum, Mohammad A Rashid, and Mohammad S Amran. 2012. "A comparative study of antidiabetic activity of hexane-extract of lemon peel (*Limon citrus*) and glimepiride in alloxan-induced diabetic rats." *Bangladesh Pharmceutical Journal* 15 (2): 131–134. doi:10.3329/BPJ.V15I2.12577.

Nair, Sunil, and John P H Wilding. 2010. "Sodium glucose cotransporter 2 inhibitors as a new treatment for diabetes mellitus." *The Journal of Clinical Endocrinology and Metabolism* 95 (1): 34–42.

Nammi, Srinivas, Murthy K Boini, Srinivas D Lodagala, and Ravindra B S Behara. 2003. "The juice of fresh leaves of *Catharanthus roseus* Linn. reduces blood glucose in normal and alloxan diabetic rabbits." *BMC Complementary and Alternative Medicine* 3 (4): 1–4. doi:10.1186/1472-6882-3-4.

National Center for Biotechnology Information. 2022. *Phenformin*. 03 10. https://pubchem.ncbi.nlm.nih.gov/compound/Phenformin#section=Toxicity.

National Institutes of Health. 2021. *Zinc: Fact Sheet for Health Professionals*. National Institutes of Health: Office of Dietary Supplements. March 26. https://ods.od.nih.gov/factsheets/Zinc-HealthProfessional/.

Nawaz, Aamir, Muhammad A Ayub, Farwa Nadeem, and Jamal N Al-Sabahi. 2016. "Lantana (*Lantana camara*): A medicinal plant having high therapeutic potentials – A comprehensive review." *International Journal of Chemical and Biochemical Sciences* 10: 52–59.

Nayak, B Shivananda, Julien R Marshall, Godwin Isitor, and Andrew Adogwa. 2011. "Hypoglycemic and hepatoprotective activity of fermented fruit juice of *Morinda citrifolia* (Noni) in diabetic rats." *Evidence-Based Complementary and Alternative Medicine* 2011: 1–5. doi:10.1155/2011/875293.

Nelson, David L, and Michael M Cox. 2008. *Lehninger Principles of Biochemistry*. 5th ed. New York: W.H. Freeman and Company.

Nelson, Kathryn M, Jayme L Dahlin, Jonathan Bisson, James Graham, Guido F Pauli, and Michael A Walters. 2017. "The essential medicinal chemistry of curcumin: Miniperspective." *Journal of Medicinal Chemistry* 60 (5): 1620–1637. doi:10.1021/acs.jmedchem.6b00975.

N'Guessan Bra Fofie, Yvette, Rokia Sanogo, Birama Diarra, Fatoumata Kanadjigui, and Dieneba Kone-Bamba. 2013. "Antioxidant and anti-hyperglycaemic activity of *Euphorbia hirta* L. on Wistar rats." *The International Journal of Biological and Chemical Sciences* 7 (6): 2558–2567. doi:10.4314/ijbcs.v7i6.30.

Niaz, Khalid, Seemi Gull, and M A Zia. 2013. "Antihyperglycemic/hypoglycemic effect of celery seeds (Ajwain/Ajmod) in streptozotocin induced diabetic rats." *Journal of Rawalpindi Medical College* 17 (1): 134–137.

Nimesh, Saurabh, Ravi Tomar, and Shubham Dhiman. 2019. "Medicinal herbal plants and allopathic drugs to treat diabetes mellitus: A glance." *Advances in Pharmacology and Clinical Trials* 4 (1): 1–13. doi:10.23880/apct-16000151.

Nordisk, Novo. 2017. "PRANDIN." 2. Accessed 03 10, 2022. https://www.novo-pi.com/prandin.pdf.

Nyunaï, Nyemb, Adèle Manguelle-Dicoum, Njikam Njifutié, El H Abdennebi, and Cros Gérard. 2010. "Antihyperglycaemic effect of *Ageratum conyzoides* L. fractions in normoglycemic and diabetic male wistar rats." *International Journal of Biomedical and Pharmaceutical Sciences* 4 (1): 38–42.

Nyunaï, Nyemb, Njifutié Njikam, El H Abdennebi, Joseph T Mbafor, and Driss Lamnaouer. 2009. "Hypoglycaemic and antihyperglycaemic activity of *Ageratum conyzoides* L. in rats." *African Journal of Traditional, Complementary and Alternative Medicines* 6 (2): 123–130.

Oguanobi, Nelson I, Chioli P Chijioke, and Samuel I Ghasi. 2012. "Effects of aqueous leaf extract of *Ocimum gratissimum* on oral glucose tolerance test in type-2 model diabetic rats." *African Journal of Pharmacy and Pharmacology* 6 (9): 630–635.

Oguanobi, Nelson I, Chioli P Chijioke, Samuel I Ghasl, Francis I Ukekwe, and Keneth I Nwadike. 2019. "Toxicity studies on crude leaf extract of *Ocimum gratissimum* in normoglycaemic and diabetic rats." *Research & Reviews: Journal of Pharmacology and Toxicological Studies* 7 (1): 1–7.

Ogugua, V N, S I Egba, E Anaduaka, and B O Ozioko. 2013. "Phytochemical analysis, anti-hyperglycaemic and antioxidant effect of the aqueous extract of *Chromolaena odorata* on alloxan induced diabetic rats." *An International Journal of Advances in Pharmaceutical Sciences* 4 (5): 970–977.

Oguntibeju, Oluwafemi O. 2019. "Hypoglycaemic and anti-diabetic activity of selected African medicinal plants." *International Journal of Physiology, Pathophysiology and Pharmacology* 11 (6): 224–237.

Ojewole, John A O. 2005. "Antinociceptive, anti-inflammatory and antidiabetic effects of *Bryophyllum pinnatum* (Crassulaceae) leaf aqueous extract." *Journal of Ethnopharmacology* 99 (1): 13–19. doi:10.1016/j.jep.2005.01.025.

Okada, H, C Kuhn, H Feillet and J-F Bach. 2010. "The 'hygiene hypothesis' for autoimmune and allergic diseases: An update." *Clinical and Experimental Immunology* 160 (1): 1–9. doi:10.1111/j.1365-2249.2010.04139.x.

Okoduwa, Stanley I R, Isamila A Umar, Dorcas B James, and Hajiya M Inuwa. 2017. "Anti-diabetic potential of *Ocimum gratissimum* leaf fractions in Fortified diet-fed streptozotocin treated Rat model of type-2 diabetes." *Medicines* 4 (73): 1–21. doi:10.3390/medicines4040073.

Okon, Uduak A, and Idorenyin U Umoren. 2017. "Comparison of antioxidant activity of insulin, *Ocimum gratissimum* L., and *Vernonia amygdalina* L. in type 1 diabetic rat model." *Journal of Integrative Medicine* 15 (4): 302–309. doi:10.1016/S2095-4964(17)60332-7.

Okur, Mehmet E, Hanefi Özbek, Derya Ç Polat, Sezen Yılmaz, and Rana Arslan. 2018. "Hypoglycemic activity of *Capparis ovata* desf. var. palaestina zoh. methanol extract." *Brazilian Journal of Pharmaceutical Sciences* 54 (3): 1–9. doi:10.1590/s2175-97902018000318031.

Okyar, Alper, Ayse Can, Nuriye Akev, Gül Baktir, and Nurhayat Sütlüpinar. 2001. "Effect of *Aloe vera* leaves on blood glucose level in type I and type II diabetic rat model." *Phytotherapy Research* 15 (2): 157–161. doi:10.1002/ptr.719.

Oladeji, Oluwole S, Funmilayo E Adelowo, David T Ayodele, and Kehinde A Odelade. 2019. "Phytochemistry and pharmacological activities of *Cymbopogon citratus*: A review." *Scientific African* 6: 1–11. doi:10.1016/j.sciaf.2019.e00137.

Oloyede, Ganiyat K, Patricia A Onocha, and Bamidele B Olaniran. 2011. "Phytochemical, toxicity, antimicrobial and antioxidant screening of leaf extracts of *Peperomia pellucida* from Nigeria." *Advances in Environmental Biology* 5 (12): 3700–3709.

Omari, Al, Somaiyah Mohammed, Abu Rjai, Talal Ahmad, and Ahmad M Disi. 2007. "The effect of low doses of myrcene and thujone and a combination of both on diabetic rats." *Dar AlMandumah*, 1–49. http://search.mandumah.com/Record/549100.

Omonije, Oluyemisi O, Abubakar N Saidu, and Hadiza L Muhammad. 2019. "Anti-diabetic activities of *Chromolaena odorata* methanol root extract and its attenuation effect on diabetic induced hepatorenal impairments in rats." *Clinical Phytoscience* 5 (23): 1–10. doi:10.1186/s40816-019-0115-1.

Omonkhua, Akhere A, and Iyere O Onoagbe. 2007. "Effects of *Irvingia grandifolia*, *Urena lobata* and *Carica papaya* on the oxidative status of normal rabbits." *The Internet Journal of Nutrition and Wellness* 6 (2): 1–11.

Omonkhua, Akhere A, and Iyere O Onoagbe. 2011. "Evaluation of the long-term effects of *Urena lobata* root extracts on blood glucose and hepatic function of normal rabbits." *Journal of Toxicology and Environmental Health Sciences* 3 (8): 204–213. doi:10.5897/JTEHS.9000031.

Omonkhua, Akhere A, and Iyere O Onoagbe. 2012. "Long-term effects of three hypoglycaemic plants (Irvingia gabonensis, *Urena lobata* and *Carica papaya*) on the oxidative status of normal rabbits." *Biokemistri* 24 (2): 82–89.

Omonkhua, Akhere A, and Iyere O Onoagbe. 2017. "Long term effect of *Urena lobata* administration on the serum lipid panel and atherogenic indices of normal rabbits." *Nigerian Journal of Pure and Applied Sciences* 30 (3): 3085–3091. doi:10.19240/njpas.2017.C10.

Onaolapo, Adejoke Y, Olakunle J Onaolapo, and Stephen O Adewole. 2012. "*Ocimum Gratissimum* Linn worsens streptozotocin-induced nephrotoxicity in diabetic wistar rats." *Macedonian Journal of Medical Sciences* 5 (4): 382–388. doi:10.3889/MJMS.1857-5773.2012.0244.

Onkaramurthy, M, V P Veerapur, B S Thippeswamy, T N M Reddy, Hunasagi Rayappa, and S Badami. 2013. "Anti-diabetic and anti-cataract effects of *Chromolaena odorata* Linn., in streptozotocin-induced diabetic rats." *Journal of Ethnopharmacology* 145: 363–372. doi:10.1016/j.jep.2012.11.023.

Östenson, C-G, A Nylén, V Grill, M Gutniak, and S Efendić. 1986. "Sulfonylurea-induced inhibition of glucagon secretion from the perfused rat pancreas: Evidence for a direct, non-paracrine effect." *Diabetologia* 29 (12): 861–867. doi:10.1007/BF00870141.

Oubré, A Y, T J Carlson, S R King, and G M Reaven. 1997. "From plant to patient: An ethnomedical approach to the identification of new drugs for the treatment of NIDDM." *Diabetologia* 40: 614–617.

Owira, P M O, and J A O Ojewole. 2009. "Grapefruit juice improves glycemic control but also exacerbates metformin-induced lactic acidosis in non-diabetic rats." *Methods & Findings in Experimental & Clinical Pharmacology* 31 (9): 563–570.

Ozcelilkay, A T, Dominique J Becker, Lumbe N Ongemba, Anne-Marie Pottier, Jean-Claude Henquin, and Sonia M Brichard. 1996. "Improvement of glucose and lipid metabolism in diabetic rats treated with molybdate." *American Journal of Physiology* 270 (2): E344–E352. doi:10.1152/ajpendo.1996.270.2.E344.

Pamplona-Roger, George D. 2005. *Encyclopedia of Medicinal Plants Education and Health Library*. Vol. II. II vols. Madrid: Editorial-Safeliz.

Pamunuwa, Geethi, D N Karunaratne, and Viduranga Y Waisundara. 2016. "Antidiabetic properties, bioactive constituents, and other therapeutic effects of *Scoparia dulcis*." *Evidence-Based Complementary and Alternative Medicine* 2016: 1–11. doi:10.1155/2016/8243215.

Panacea Biotec Ltd. 2019. "VILACT." *Vilact*. December. https://vilact.panaceabiotec.com/assets/documents/Vilact-Product-Monograph.pdf.

Pandey, Kirti. 2020. "Neem extracts: Treatment of diabetes mellitus gains fillip from use of neem leaves to lower blood sugar levels." *TimesNowNews.com*. September 23. Accessed July 22, 2021. https://www.timesnownews.com/health/article/neem-extracts-treatment-of-diabetes-mellitus-gains-fillip-from-use-of-neem-leaves-to-lower-blood-sugar-levels/656609.

Panneerselvam, Saraswathi R, and Swaminathan Govindasamy. 2004. "Effect of sodium molybdate on the status of lipids, lipid peroxidation and antioxidant systems in alloxan-induced diabetic rats." *Clinica Chimica Acta* 345 (1–2): 93–98. doi:10.1016/j.cccn.2004.03.005.

Parasuraman, Subramani, Subramani Balamurugan, Parayil V Christapher, Rajendran R Petchi, Wong Y Yeng, Jeyabalan Sujithra, and Chockalingam Vijaya. 2015. "Evaluation of antidiabetic and antihyperlipidemic effects of hydroalcoholic extract of leaves of *Ocimum tenuiflorum* (Lamiaceae) and prediction of biological activity of its phytoconstituents." *Pharmacognosy Research* 7 (2): 156–165. doi:10.4103/0974-8490.151457.

Paredi, P, G Invernizzi W Biernacki, S A Kharitonov, and P J Barnes. 1999. "Exhaled carbon monoxide levels elevated in diabetes and correlated with glucose concentration in blood: A new test for monitoring the disease?" *Chest* 116 (4): 1007–1011.

Pari, Leelavinothan, and Muniappan Latha. 2004. "Antihyperglycaemic effect of *Scoparia dulcis*: Effect on key metabolic enzymes of carbohydrate metabolism in streptozotocin-induced diabetes." *Pharmaceutical Biology* 42 (8): 570–576. doi:10.1080/13880200490901799.

Pari, Leelavinothan, and Pidaran Murugan. 2007. "Tetrahydrocurcumin prevents brain lipid peroxidation in streptozotocin-induced diabetic rats." *Journal of Medicinal Food* 10 (2): 323–329. doi:10.1089/jmf.2006.058.

Parke Davis Pharmaceuticals Limited. 1999. "REZULIN." 06. Accessed 03 10, 2022. https://www.accessdata.fda.gov/drugsatfda_docs/label/1999/20720s12lbl.pdf.

Parmar, Hamendra S, and Anand Kar. 2007. "Antidiabetic potential of *Citrus sinensis* and Punica granatum peel extracts in alloxan treated male mice." *BioFactors* 31 (1): 17–24. doi:10.1002/biof.5520310102.

Parmar, Hamendra S, and Anand Kar. 2008a. "Antiperoxidative, antithyroidal, antihyperglycemic and cardioprotective role of *Citrus sinensis* peel extract in male mice." *Phytotherapy Research* 22 (6): 791–795. doi:10.1002/ptr.2367.

Parmar, Hamendra S, and Anand Kar. 2008b. "Medicinal values of fruit peels from *Citrus sinensis*, *Punica granatum*, and *Musa paradisiaca* with respect to alterations in tissue lipid peroxidation and serum concentration of glucose, insulin, and thyroid hormones." *Journal of Medicinal Food* 11 (2): 376–381. doi:10.1089/jmf.2006.010.

Parvin, Anzana, Morshedul Alam, Anwarul Haque, Amrita Bhowmik, Liaquat Ali, and Begum Rokeya. 2013. "Study of the hypoglycemic effect of *Tamarindus indica* Linn. seeds on non-diabetic and diabetic model rats." *British Journal of Pharmaceutical Research* 3 (4): 1094–1105. doi:10.9734/BJPR/2013/4865.

Patel, Vijay S, Vellapandian Chitra, P L Prasanna, and V Krishnaraju. 2008. "Hypoglycemic effect of aqueous extract of *Parthenium hysterophorus* L. in normal and alloxan induced diabetic rats." *Indian Journal of Pharmacology* 40 (4): 183–185.

Patumraj, Suthiluk, Natchaya Wongeakin, Patarin Sridulyakul, Amporn Jariyapongskul, Narisa Futrakul, and Srichitra Bunnag. 2006. "Combined effects of curcumin and vitamin C to protect endothelial dysfunction in the iris tissue of STZ-induced diabetic rats." *Clinical Hemorheology and Microcirculation* 35 (4): 481–489.

Patumraj, Suthiluk, S Tewit, S Amatyakul, Amporn Jaryiapongskul, Sangrawee Maneesri, V Kasantikul, and D Shepro. 2000. "Comparative effects of garlic and aspirin on diabetic cardiovascular complications." *Drug Delivery* 7 (2): 91–96. doi:10.1080/107175400266650.

Paudyal, Buddhi, Astha Thapa, Keshav R Sigdel, Sudeep Adhikari, and Buddha Basnyat. 2019. "Adverse events with ayurvedic medicines- possible adulteration and some inherent toxicities." *Wellcome Open Access* 4 (23): 1–15.

Paul, Meera, Kavitha Vasudevan, and K R Krishnaja. 2017. "*Scoparia dulcis*: A review on its phytochemical and pharmacological profile." *Innoriginal International Journal of Sciences* 4 (4): 17–21.

Paula, Paulo, Sousa Daniele, Jose Oliviera, Ana Carvalho, Bella Alves, Mirella Periera, Davi Farias, et al. 2017. "A protein isolate from *Moringa oleifera* leaves has hypoglycemic and antioxidant effects in alloxan-induced diabetic mice." *Molecules* 22 (2): 1–15. doi:10.3390/molecules22020271.

Peng, Mei, and Xiaoping Yang. 2015. "Controlling diabetes by chromium complexes: The role of the ligands." *Journal of Inorganic Biochemistry* 146: 97–103. doi:10.1016/j.jinorgbio.2015.01.002.

Pérez-Torres, Israel, Angélica Ruiz-Ramírez, Guadalupe Baños, and Mohammed El-Hafidi. 2013. "Hibiscus sabdariffa Linnaeus (Malvaceae), curcumin and resveratrol as alternative medicinal agents against metabolic syndrome." *Cardiovascular & Hematological Agents in Medicinal Chemistry* 11 (1): 25–37. doi:10.2174/1871525711311010006.

Perry, J R B, L Ferrucci, S Bandinelli, J Guralnik, R D Semba, N Rice, D Melzer, et al. 2009. "Circulating β-carotene levels and type 2 diabetes—cause or effect?" *Diabetologia* 52 (10): 2117–2121. doi:10.1007/s00125-009-1475-8.

Peter, K V. 2012. *Handbook of Herbs and Spices*. 2nd ed. Edited by K V Peter. Vol. II. Oxford: Woodhead Publishing Limited.

Pfizer-Roerig. 2008. "GLUCOTROL." Accessed 10 03, 2022. https://www.accessdata.fda.gov/drugsatfda_docs/label/2008/017783s019lbl.pdf.

Pillai, Sneha S, and S Mini. 2016. "*Hibiscus rosa sinensis* Linn. petals modulates glycogen metabolism and glucose homeostasis signalling pathway in streptozotocin-induced experimental diabetes." *Plant Foods for Human Nutrition* 71 (1): 42–48. doi:10.1007/s11130-015-0521-6.

Pillay, P P, C P M Nair, and T N Santi Kumari. 1959. "*Lochnera rosea* as a potential source of hypotensive and other remedies." *Bulletin of Research Institute of the University of Kerala* 1: 51–54.

Pollak, Michael. 2013. "Potential applications for biguanides in oncology." *The Journal of Clinical Investigation* 123 (9): 3693–3700. doi:10.1172/JCI67232.

Pothuraju, Ramesh, Raj K Sharma, Suneel K Onteru, Satvinder Singh, and Shaik A Hussain. 2016. "Hypoglycemic and hypolipidemic effects of *Aloe vera* extract preparations: A review." *Phytotherapy Research* 30 (2): 200–207. doi:10.1002/ptr.5532.

Pramanik, Jajati, and Indrajit Giri. 2018. "Targeted anti diabetic activity of *Lantana camara*: A review." *International Journal of Botany Studies* 3 (4): 9–11.

Prashanth, D, Agarwal Amit, D S Samiulla, M K Asha, and R Padmaja. 2001. "α-Glucosidase inhibitory activity of *Mangifera indica* bark." *Fitoterapia* 72 (6): 686–688. doi:10.1016/s0367-326x(01)00293-3.

Purnomo, Yadi, Djoko W Soeatmadji, Sutiman B Sumitro, and Mochamad A Widodo. 2017. "Incretin effect of *Urena lobata* leaves extract on structure and function of rats islet beta-cells." *Journal of Traditional and Complementary Medicine* 7 (3): 301–306. doi:10.1016/j.jtcme.2016.10.001.

Purnomo, Yudi, Djoko W Soeatmadji, Sutiman B Sumitro, and Mochamad A Widodo. 2015a. "Anti-diabetic potential of *Urena lobata* leaf extract through inhibition of dipeptidyl-peptidase IV activity." *Asian Pacific Journal of Tropical Biomedicine* 5 (8): 645–649. doi:10.1016/j.apjtb.2015.05.014.

Purnomo, Yudi, Djoko W Soeatmadji, Sutiman B Sumitro, and Mochamad A Widodo. 2015b. "Anti-hiperglycemic effect of *Urena lobata* leaf extract by inhibition of dipeptidyl peptidase IV (DPP-IV) on diabetic rats." *International Journal of Pharmacognosy and Phytochemical Research* 7 (5): 1073–1079.

Purnomo, Yudi, Djoko W Soeatmadji, Sutiman B Sumitro, and Mochamad A Widodo. 2018. "Inhibitory activity of *Urena lobata* leaf extract on dipeptidyl peptidase-4 (DPP-4): Is it different in vitro and in vivo?" *Medicinal Plants - International Journal of Phytomedicines and Related Industries* 10 (2): 99–105. doi:10.5958/0975-6892.2018.00016.3.

Rabasa-Lhoret, Rémi, and Jean-Louis Chiasson. 1998. "Potential of α-glucosidase inhibitors in elderly patients with diabetes mellitus and impaired glucose tolerance." *Drugs & Aging* 13 (2): 131–143. doi:10.2165/00002512-199813020-00005.

Rafique, Sania, Syeda M Hassan, Shahzad S Mughal, Syed K Hassan, Nageena Shabbir, Sumaira Perveiz, Maryam Mushtaq, and Muhammad Farman. 2020. "Biological attributes of lemon: A review." *Journal of Addiction Medicine and Therapeutic Science* 6 (1): 030–034. doi:10.17352/2455-3484.000034.

Raghavendra, H L, and T R Prashith Kekuda. 2018. "Ethnobotanical uses, phytochemistry and pharmacological activities of *Peperomia pellucida* (L.) Kunth (Piperaceae) - A review." *International Journal of Pharmacy and Pharmaceutical Sciences* 10 (2): 1–8. doi:10.22159/ijpps.2018v10i2.23417.

Rahimi, Maryam. 2015. "A review: Anti diabetic medicinal plants used for diabetes mellitus." *Bulletin of Environment, Pharmacology and Life Sciences* 4 (2): 163–180. doi:10.1016/S2221-6189(13)60126-2.

Rajasekaran, S, K Sivagnanam, Kasiappan Ravi, and Saju Subramanian. 2004. "Hypoglycemic effect on *Aloe vera* gel on streptozotocin-induced diabetes in experimental rats." *Journal of Medicinal Food* 7 (1): 61–66. doi:10.1089/109662004322984725.

Rajendiran, Deepa, Selvakumar Kandaswamy, Senthilkumar Sivanesan, Saravanan Radhakrishnan, and Krishnamoorthy Gunasekaran. 2017. "Potential antidiabetic effect of *Mimosa pudica* leaves extract in high fat diet and low dose streptozotocin-induced type 2 diabetic rats." *International Journal of Biology Research* 2 (4): 55–62.

Ramachandran, S, V S Sandeep, N K Srinivas, and M D Dhanaraju. 2010. "Anti-diabetic activity of *Abelmoschus esculentus* Linn. on alloxan induced diabetic rats." *Research & Reviews in BioSciences* 4 (3): 121–123.

Ramasamy, Ravichandran, and Ann Marie Schmidt. 2014. "Etiology of diabetes mellitus." Vol. 12, in Ira Lamster *Diabetes Mellitus and Oral Health: An Interprofessional Approach*, 1–26. Hoboken, NJ: John Wiley & Sons, Inc.

Rambaran, Varma H, S M Saumya, Soumyabrata Roy, K P Sonu, Muthusamy Eswaramoorthy, and Sebastian C Peter. 2020. "The design, synthesis and in vivo biological evaluations of [V (IV) O (2, 6-pyridine diacetatato)(H2O) 2](PDOV): Featuring its prolonged glucose lowering effect and non-toxic nature." *Inorganica Chimica Acta* 504 (20): 119448. doi:10.1016/j.ica.2020.119448.

Rambaran, Varma, and Saumyua S. Mani. 2021. "Vanadium insulin-mimetics, methods of preparation, and methods for treatment of diabetes." *United States Patent 10953042*. March 23.

Rani, Manjeet P, K P Padmakumari, B Sankarikutty, Lijo O Cherian, V M Nisha, and K G Raghu. 2011. "Inhibitory potential of ginger extracts against enzymes linked to type 2 diabetes, inflammation and induced oxidative stress." *International Journal of Food Sciences and Nutrition* 62 (2): 106–110. doi:10.3109/09637486.2010.515565.

Raphael, K R, M C Sabu, and Ramadasan Kuttan. 2002. "Hypoglycemic effect of methanol extract of *Phyllanthus amarus* Schum & Thonn on alloxan induced diabetes mellitus in rats and its relation with antoixidant potential." *Indian Journal of Experimental Biology* 40 (8): 905–909.

Reaven, Gerald. 2004. "The metabolic syndrome or the insulin resistance syndrome? Different names, different concepts, and diffhe metaboli." *Endocrinology and Metabolic Clinincs in North America* 33: 283–303.

Reed, Michael J, and Karen A Scribner. 1999. "In-vivo and in-vitro models of type 2 diabetes in pharmaceutical drug discovery." *Diabetes, Obesity and Metabolism* 1: 75–86.

Rena, Graham, D G Hardie, and Ewan R Pearson. 2017. "The mechanisms of action of metformin." *Diabetologia* 60 (9): 1577–1585. doi:10.1007/s00125-017-4342-z.

Riaz, Azra, Rafeeq A Khan, and Mansoor Ahmed. 2013. "Glycemic response of citrus limon, pomegranate and their combinations in alloxan-induced diabetic rats." *Australian Journal of Basic and Applied Sciences* 7 (10): 215–219.

Riddle, Matthew C. 2017. "Modern sulfonylureas: Dangerous or wrongly accused?" *Diabetes Care* 40 (5): 629–631. doi:10.2337/dci17-0003.

Riddle, Matthew C, and William T Cefalu. 2018. "SGLT inhibitors for type 1 diabetes: An obvious choice or too good to be true?" *Diabetes Care* 41 (12): 2444–2447. doi:10.2337/dci18-0041.

Rosalie, Ijomone O, and E L Ekpe. 2016. "Antidiabetic potentials of common herbal plants and plant products: A glance." *International Journal of Herbal Medicine* 4 (4): 90–97.

Rosenstock, Julio, David R Hassman, Robert D Madder, Shari A Brazinsky, James Farrell, Naum Khutoryansky, and Paula M Hale. 2004. "Repaglinide versus nateglinide monotherapy: A randomized, multicenter study." *Diabetes Care* 27 (6): 1265–1270. doi:10.2337/diacare.27.6.1265.

Ross, Ivan A. 2003. *Medicinal Plants of the World: Chemical Constituents,Traditional and Modern Medicinal Uses.* 2nd. Vol. I. III vols. New York: Springer Science+Business Media.

Ross, Ivan A. 2005. *Medicinal Plants of the World: Chemical Constituents, Traditional and Modern Medicinal Uses.* Vol. III. III vols. Totowa, NJ: Humana Press.

Rowles, Alexandra. 2017. "Why molybdenum is an essential nutrient." *Healthline.* May 6. Accessed 8 28, 2021. https://askinglot.com/open-detail/105242.

Rozianoor, M H W, Y N Eizzatie, and Samsulrizal Nurdiana. 2014. "Hypoglycemic and antioxidant activities of *Stachytarpheta jamaicensis* ethanolic leaves extract on alloxan-induced diabetic sprague dawley rats." *BioTechnology: An Indian Journal* 9 (10): 423–428.

Russell, K R M, E Y St A Morrison, and D Ragoobirsingh. 2005. "The effect of annatto on insulin binding properties in the dog." *Phytotherapy Research* 19 (5): 433–436. doi:10.1002/ptr.1650.

Sabitha, V, S Ramachandran, K R Naveen, and K Panneerselvam. 2011. "Antidiabetic and antihyperlipidemic potential of *Abelmoschus esculentus* (L.) Moench. in streptozotocin-induced diabetic rats." *Journal of Pharmacy and Bioallied Sciences* 3 (3): 397–402. doi:10.4103/0975-7406.84447.

Sachdewa, Archana, and L D Khemani. 2003. "Effect of *Hibiscus rosa sinensis* Linn. ethanol flower extract on blood glucose and lipid profile in streptozotocin induced diabetes in rats." *Journal of Ethnopharmacology* 89 (1): 61–66. doi:10.1016/s0378-8741(03)00230-7.

Sachdewa, Archana, Rashmi Nigam, and L D Khemani. 2001. "Hypoglycemic effect of *Hibiscus rosa sinensis* L. leaf extract in glucose and streptozotocin induced hyperglycemic rats." *Indian Journal of Experimental Biology* 39 (3): 284–286.

Sahebkar, Amirhossein. 2013. "Why it is necessary to translate curcumin into clinical practice for the prevention and treatment of metabolic syndrome?" *Biofactors* 39 (2): 197–208. doi:10.1002/biof.1062.

Sahrawat, Alka, Jyoti Sharma, Siddarth N Rahul, Snigdha Tiwari, and D V Rai. 2018. "*Parthenium hysterophorus* current status and its possible effects on mammalians - A review." *International Journal of Current Microbiology and Applied Sciences* 7 (11): 3548–3557.

Saito, Mitsuo, Mutsuko Hirata-Koizumi, Mariko Matsumoto, Tsutomu Urano, and Ryuichi Hasegawa. 2005. "Undesirable effects of citrus juice on the pharmacokinetics of drugs: Focus on recent studies." *Drug Safety* 28 (8): 677–694. doi:10.2165/00002018-200528080-00003.

Saker, Firas, Juan Ybarra, Patrick Leahy, Richard W Hanson, Satish C Kalhan, and Faramarz Ismail-Beigi. 1998. "Glycemia-lowering effect of cobalt chloride in the diabetic rat: Role of decreased gluconeogenesis." *American Journal of Physiology - Endocrinology and Metabolism* 274 (6): E984–E991. doi:10.1152/ajpendo.1998.274.6.E984.

Saleem, Uzma, Muhammad Ejaz-ul-Haq, Zunera Chudary, and Bashir Ahmad. 2017. "Pharmacological screening of *Annona muricata*: A review." *Asian Journal of Agriculture and Biology* 5 (1): 38–46.

Sánchez, G M, L Re, A Giuliani, A J Núñez-Sellés, G P Davison, and O S León-Fernández. 2000. "Protective effects of *Mangifera indica* L. extract, mangiferin and selected antioxidants against TPA-induced biomolecules oxidation and peritoneal macrophage activation in mice." *Pharmacological Research* 42 (6): 565–573. doi:10.1006/phrs.2000.0727.

Sangeetha, M, C Mahendran, and C Ushadevi. 2015. "An overview on the medicinal properties of *Lantana camara* Linn." *International Journal of Innovative Pharmaceutical Sciences and Research* 3 (5): 645–654.

Sanofi-Aventis. 2009. "DIABETA." Accessed 03 10, 2022. https://www.accessdata.fda.gov/drugsatfda_docs/label/2009/017532s030lbl.pdf.

Sanofi-Aventis. 2013. "AMARYL." Accessed 10 03, 2022. https://www.accessdata.fda.gov/drugsatfda_docs/label/2013/020496s027lbl.pdf.

Sarkar, Ahana, Pranabesh Ghosh, Susmita Poddar, Tanusree Sarkar, Suradipa Choudhury, and Sirshendu Chatterjee. 2020. "Phytochemical, botanical and ethnopharmacological study of *Scoparia dulcis* Linn. (Scrophulariaceae): A concise review." *The Pharma Innovation Journal* 9 (7): 30–35. doi:10.22271/tpi.2020.v9.i7a.5049.

Sarkar, Shubhashish, Maddali Pranava, and Rosalind A Marita. 1996. "Demonstration of the hypoglycemic action of *Momordica charantia* in a validated animal model of diabetes." *Pharmacological Research* 33 (1): 1–4. doi:10.1006/phrs.1996.0001.

Sathiyabama, Rajiv, Gopalsamy Gandhi, Marina Denadai, Gurunagarajan Sridharan, Gnanasekaran Jothi, Ponnusamy Sasikumar, Jullyana Quintans, et al. 2018. "Evidence of insulin-dependent signalling mechanisms produced by *Citrus sinensis* (L.) Osbeck fruit peel in an insulin resistant diabetic animal model." *Food and Chemical Toxicology* 116: 86–99. doi:10.1016/j.fct.2018.03.050.

Satyanarayana, K, K Sravanthi, I A Shaker, and R Ponnulakshmi. 2015. "Molecular approach to identify antidiabetic potential of *Azadirachta indica*." *Journal of Ayurveda & Integrative Medicine* 6 (3): 165–174. doi:10.4103/0975-9476.157950.

Schulz, Volker, Rudolf Hänsel, Mark Blumenthal, and Varro E Tyler. 2004. "Agents that increase resistance to diseases." Chap. 9 in *Rational Phytotherapy: A Physicians' Guide to Herbal Medicine*, 369–398. Berlin: Springer. doi:10.1007/978-3-662-09666-6_9.

Scior, Thomas, Jose A Guevara-Garcia, Quoc-Tuan Do, Philippe Bernard, and Stefan Laufer. 2016. "Why antidiabetic vanadium complexes are not in the pipeline of 'Big Pharma' drug research? A critical review." *Current Medicinal Chemistry* 23 (25): 2874–2891. doi:10.2174/0929867323666160321121138.

Scott, Lesley J, and Caroline M Spencer. 2000. "Miglitol: A review of its therapeutic potential in type 2 diabetes mellitus." *Drugs* 59 (3): 521–549. doi:10.2165/00003495-200059030-00012.

Selvanathan, Jaganathan, and Subramani Sundaresan. 2020. "Acute toxicity effects on antidiabetic activities of *Chromolaena odorata* in induced wistar albino rats." *A Journal of Composition Theory* 13 (1): 206–222.

Servier Canada Inc. 2018. "DIAMICRON MR." *Product Monograph: Diamicron*. 01 02. https://pdf.hres.ca/dpd_pm/00043004.PDF.

Sesti, Giorgio. 2006. "Pathophysiology of insulin resistance." *Best Practice & Research Clinical Endocrinology & Metabolism* 20 (4): 665–679. doi:10.1016/j.beem.2006.09.007.

Shah, Gagan, Richa Shri, Vivek Panchal, Narender Sharma, Bharpur Singh, and A S Mann. 2011. "Scientific basis for the therapeutic use of *Cymbopogon citratus*, Stapf (Lemon grass)." *Journal of Advanced Pharmaceutical Technology & Research* 2 (1): 3–8. doi:10.4103/2231-4040.79796.

Shah, K A, M B Patel, R J Patel, and P K Parmar. 2010. "*Mangifera indica* (mango)." *Pharmacognosy Reviews* 4 (7): 42–48. doi:10.4103/0973-7847.65325.

Shakib, Zahra, Naghmeh Shahraki, Bibi M Razavi, and Hossein Hosseinzadeh. 2019. "*Aloe vera* as an herbal medicine in the treatment of metabolic syndrome: A review." *Phytotherapy Research* 33 (4): 2649–2660. doi:10.1002/ptr.6465.

Sharma, Ajay K, and Ritika Gupta. 2017. "Anti-hyperglycemic activity of aqueous extracts of some medicinal plants on wistar rats." *Journal of Diabetes and Metabolism* 8 (7): 1–7. doi:10.4172/2155-6156.1000752.

Sharma, Ashok K, Saurabh Bharti, Rajiv Kumar, Bhaskar Krishnamurthy, Jagriti Bhatia, Santosh Kumari, and Dharamvir S Arya. 2012. "*Syzygium cumini* ameliorates insulin resistance and β-cell dysfunction via modulation of PPARγ, dyslipidemia, oxidative stress, and TNF-α in type 2 diabetic rats." *Journal of Pharmacological Sciences* 119 (3): 205–213. doi:10.1254/jphs.11184fp.

Sharma, Gunjan, Umama Ashhar, Vidhu Aeri, and Deepshikha P Katare. 2018. "Effect of ethanolic extract of *Euphorbia hirta* on chronic diabetes mellitus and associated cardiorenal damage in rats." *International Journal of Green Pharmacy* 12 (3): 191–199.

Sharma, K K, R K Gupta, S Gupta, and K C Samuel. 1977. "Antihyperglycemic effect of onion: Effect on fasting blood sugar and induced hyperglycemia in man." *Indian Journal of Medical Research* 65 (4): 422–429.

Sharmin, Rifath, M R I Khan, M A Akhtar, A Alim, M A Islam, Abu S M Anisuzzaman, and Maisa Ahmed. 2013. "Hypoglycemic and hypolipidemic effects of cucumber, white pumpkin and ridge gourd in alloxan induced diabetic rats." *Journal of Scientific Research* 5 (1): 161–170. doi:10.3329/jsr.v5i1.10252.

Sheela, C G, and K T Augusti. 1992. "Antidiabetic effects of S-allyl cysteine sulphoxide isolated from garlic Allium sativum Linn." *Indian Journal of Experimental Biology* 30 (6): 523–526.

Sheela, C G, K Kumud, and K T Augusti. 1995. "Anti-diabetic effects of onion and garlic sulfoxide amino acids in rats." *Planta Medica* 61 (4): 356–357.

Shehzad, Adeeb, Taewook Ha, Fazli Subhan, and Young S Lee. 2011. "New mechanisms and the anti-inflammatory role of curcumin in obesity and obesity-related metabolic diseases." *European Journal of Nutrition* 50 (3): 151–161. doi:10.1007/s00394-011-0188-1.

Sheikh, Hasib, Shotabdi Sikder, Sagar K Paul, A M R Hasan, Mofizur Rahaman, and Sangita P Kundu. 2013. "Hypoglycemic, anti-inflammatory and analgesic activity of *Peperomia pellucida* (L.) HBK (Piperaceae)." *International Journal of Pharmaceutical Sciences and Research* 4 (1): 458–463.

Shetti, A A, R D Sanakal, and B B Kaliwal. 2012. "Antidiabetic effect of ethanolic leaf extract of *Phyllanthus amarus* in alloxan induced diabetic mice." *Asian Journal of Plant Science and Research* 2 (1): 11–15.

Shi, Yu, Chao Zhang, and Xiaodong Li. 2021. "Traditional medicine in India." *Journal of Traditional Chinese Medical Sciences* 8: S51–S55. doi:10.1016/j.jtcms.2020.06.007.

Shilpa, V S, S Lekshmi, and T S Swapna. 2020. "In vitro antidiabetic potential of *Euphorbia hirta* Linn.: A nutritionally significant plant." *Journal of Pharmacognosy and Phytochemistry* 9 (1): 1–4.

Singh, Som N, Praveen Vats, Shoba Suri, Radhey Shyam, M M L Kumria, S Ranganathan, and K Sridharan. 2001. "Effect of an antidiabetic extract of *Catharanthus roseus* on enzymic activities in streptozotocin induced diabetic rats." *Journal of Ethnopharmacology* 76 (3): 269–277. doi:10.1016/s0378-8741(01)00254-9.

Soetikno, Vivian, Flori R Sari, Punniyakoti T Veeraveedu, Rajarajan A Thandavaryan, Meilei Harima, Vijayakumar Sukumaran, Arun P Lakshmanan, Kenji Suzuki, Hiroshi Kawchi, and Kenichi Watanabe. 2011. "Curcumin ameliorates macrophage infiltration by inhibiting NF-κB activation and proinflammatory cytokines in streptozotocin induced-diabetic nephropathy." *Nutrition & Metabolism* 8 (1): 35. doi:10.1186/1743-7075-8-35.

Soetikno, Vivian, Kenichi Watanabe, Flori R Sari, Meilei Harima, Rajarajan A Thandavarayan, Punniyakoti T Veeraveedu, Wawaimuli Arozal, et al. 2011. "Curcumin attenuates diabetic nephropathy by inhibiting PKC-α and PKC-β1 activity in streptozotocin-induced type I diabetic rats." *Molecular Nutrition & Food Research* 55 (11): 1655–1665. doi:10.1002/mnfr.201100080.

Sola, Daniele, Luca Rossi, Gian P C Schianca, Pamela Maffioli, Marcello Bigliocca, Roberto Mella, Francesca Corlianò, Gian P Fra, Ettore Bartoli, and Giuseppe Derosa. 2015. "Sulfonylureas and their use in clinical practice." *Archives of Medical Science* 11 (4): 840–848. doi:10.5114/aoms.2015.53304.

Sotaniemi, Eero A, Eila Haapakoski, and Arja Rautio. 1995. "Ginseng therapy in non-Insulin-dependent diabetic patients." *Diabetes Care* 18 (10): 1373–1375. doi:10.2337/diacare.18.10.1373.

Srivastava, Y, H Venkatakrishna-Bhatt, Y Verma, K Venkaiah, and B H Raval. 1993. "Antidiabetic and adaptogenic properties of *Momordica charantia* extract: An experimental and clinical evaluation." *Phytotherapy Research* 7 (4): 285–289. doi:10.1002/ptr.2650070405.

Srivastava, Y, H Venkatarishna-Bhatt, O Gupta, and P Gupta. 1983. "Hypoglycemia induced by *Syzygium cumini* linn. seeds in diabetes mellitus." *Asian Medical Journal* 26 (7): 489–491.

Staff, DNT. 2018. *Vanadium Supplements for Diabetes.* Diabetes Talk. January 06. Accessed 03 01, 2022. https://diabetestalk.net/diabetes/vanadium-supplements-for-diabetes.

Subramanian, Sorimuthu P, Subramanian Bhuvaneshwari, and Gopalan S Prasath. 2011. "Antidiabetic and antioxidant potentials of *Euphorbia hirta* leaves extract studied in streptozotocin-induced experimental diabetes in rats." *General Physiology and Biophysics* 30 (3): 278–285. doi:10.4149/gpb_2011_03_278.

Sundaram, Ramalingam, Palanivelu Shanthi, and Panchanatham Sachdanandam. 2013. "Effect of iridoid glucoside on plasma lipid profile, tissue fatty acid changes, inflammatory cytokines, and GLUT4 expression in skeletal muscle of streptozotocin-induced diabetic rats." *Molecular and Cellular Biochemistry* 380 (1–2): 43–55. doi:10.1007/s11010-013-1656-0.

Sundaresan, Arjunan, and Thangaiyan Radhiga. 2015. "Effect of *Mimosa pudica* cured extract against high fructose diet induced type 2 diabetes in rats." *International Letters of Natural Sciences* 39: 1–9.

Susilawati, Yasmiwar, Ricky Nugraha, Jegatheswaran Krishnan, Ahmad Muhtadi, Supriyatna Sutardjo, and Unang Supratman. 2017. "A new antidiabetic compound 8,9-dimethoxy ellagic acid from sasaladaan (*Peperomia pellucida* L. Kunth)." *Research Journal of Pharmaceutical, Biological and Chemical Sciences* 8 (1S): 269–274.

Sutar, Nitin G, U N Sutar, and B C Behera. 2009. "Antidiabetic activity of the leaves of *Mimosa pudica* Linn. in albino rats." *Journal of Herbal Medicine and Toxicology* 3 (1): 123–126.

Szollosi, Reka. 2011. "Indian mustard (*Brassica juncea* L.) seeds in health." Chap. 78 in *Nuts and Seeds in Health and Disease Prevention*, edited by Victor Preedy and Ronald Watson, 671–676. Academic Press. doi:10.1016/B978-0-12-375688-6.10078-7.

Tachibana, Yukari, Hiroe Kikuzaki, Nordin H Lajis, and Nobuji Nakatani. 2001. "Antioxidative activity of carbazoles from *Murraya koenigii* leaves." *Journal of Agricultural and Food Chemistry* 49 (11): 5589–5594. doi:10.1021/jf010621r.

Takeda Pharmaceuticals America, Inc. 2011. "ACTOS." 07. Accessed 03 10, 2022. https://www.accessdata.fda.gov/drugsatfda_docs/label/2011/021073s043s044lbl.pdf.

Talba, Tahirou, Xia W Shui, Qinyuan Cheng, and Xin Tian. 2011. "Antidiabetic effect of glucosaminic acid-cobalt (II) chelate in streptozotocin-induced diabetes in mice." *Diabetes, Metabolic Syndrome and Obesity: Targets and Therapy* 4: 137–140. doi:10.2147/DMSO.S18025.

Tang, Yujiao, Eun-Ju Choi, Weon C Han, Mirae Oh, Jin Kim, Ji-Young Hwang, Pyo-Jam Park, Sang-Ho Moon, Yon-Suk Kim, and Eun-Kyung Kim. 2017. "*Moringa oleifera* from cambodia ameliorates oxidative stress, hyperglycemia, and kidney dysfunction in type 2 diabetic mice." *Journal of Medicinal Food* 20 (5): 502–510. doi:10.1089/jmf.2016.3792.

Teixeira, Claudio C, Letícia S Weinert, Daniel C Barbosa, Cristina Ricken, Jorge F Esteves, and Flávio D Fuchs. 2005. "*Syzygium cumini* (L.) skeels in the treatment of type 2 diabetes: Results of a randomized, double-blind, double-dummy, controlled trial." *Diabetes Care* 27 (12): 3019–3020.

Teles, Flávio, Felipe Silveira dos Anjos, Tarcilo Machado, and Roberta Lima. 2014. "*Bixa orellana* (annatto) exerts a sustained hypoglycemic effect in experimental diabetes mellitus in rats." *Medical Express* 1 (1): 36–38. doi:10.5935/MedicalExpress.2014.01.08.

Thanakosai, Wannisa, and Preecha Phuwapraisirisan. 2013. "First identification of α-glucosidase inhibitors from okra (*Abelmoschus esculentus*) seeds." *Natural Product Communications* 8 (8): 1085–1088. doi:10.1177/1934578X1300800813.

The Environmental Weeds of Australia. n.d. *Ageratum conyzoides (Billygoat Weed): BioNET-EAFRINET.* Accessed August 14, 2020. https://keys.lucidcentral.org/keys/v3/eafrinet/weeds/key/weeds/Media/Html/Ageratum_conyzoides_(Billygoat_Weed).htm.

Theiler, Barbara A, Stefanie Istvanits, Martin Zehl, Laurence Marcourt, Ernst Urban, Lugardo O E Caisa, and Sabine Glasl. 2016. "HPTLC bioautography guided isolation of α-glucosidase inhibiting compounds from *Justicia secunda* Vahl (Acanthaceae)." *Phytochemical Analysis* 28 (2): 87–92. doi:10.1002/pca.2651.

Thirumalai, T, S V Therasa, E K Elumalai, and E David. 2011. "Hypoglycemic effect of *Brassica juncea* (seeds) on streptozotocin induced diabetic male albino rat." *Asian Pacific Journal of Tropical Biomedicine* 1 (4): 323–325. doi:10.1016/S2221-1691(11)60052-X.

Thomas, Gareth. 2007. *Medicinal Chemistry: An Introduction*. 2nd ed. Chichester, West Sussex: John Wiley & Sons Ltd.

Thompson, Katherine H, and Chris Orvig. 2006. "Vanadium in diabetes: 100 years from phase 0 to phase I." *Journal of Inorganic Biochemistry* 100 (12): 1925–1935. doi:10.1016/j.jinorgbio.2006.08.016.

Thompson, Katherine H, Barry Liboiron, Yan Sun, Karycia Bellman, Ika Setyawati, Brian Patrick, Veranja Karunaratne, et al. 2003. "Preparation and characterization of vanadyl complexes with bidentate maltol-type ligands; in vivo comparisons of anti-diabetic therapeutic potential." *Journal of Biological Inorganic Chemistry* 8 (1): 66–74. doi:10.1007/s00775-002-0388-5.

Thompson, Katherine H, Jay Lichter, Carl LeBel, Michael C Scaife, John H McNeill, and Chris Orvig. 2009. "Vanadium treatment of type 2 diabetes: A view to the future." *Journal of Inorganic Biochemistry* 103 (4): 554–558. doi:10.1016/j.jinorgbio.2008.12.003.

Towers, G H N, and P V Subba Rao. 1992. "Impact of the pan-tropical weed, *Parthenium hysterophorus* L. on human affairs." First International Weed Control Congress, Melbourne, 134–138.

TROPILAB® INC. n.d. *Leonotis Nepetifolia - Lion's Ear.* TROPILAB® INC. Accessed October 20, 2020. http://www.tropilab.com/lionsear.html.

Tunna, Tasnuva S, Islam S M Zaidul, Qamar U Ahmed, Kashif Ghafoor, Fahad Y Al-Juhaimi, M S Uddin, M Hasan, and S Ferdous. 2015. "Analyses and profiling of extract and fractions of neglected weed *Mimosa pudica* Linn. traditionally used in Southeast Asia to treat diabetes." *South African Journal of Botancy* 99: 144–152. doi:10.1016/j.sajb.2015.02.016.

Tyagi, Sandeep, Paras Gupta, Arminder S Saini, Chaitnya Kaushal, and Saurabh Sharma. 2011. "The peroxisome proliferator-activated receptor: A family of nuclear receptors role in various diseases." *Journal of Advanced Pharmaceutical Technology & Research* 2 (4): 236–240. doi:10.4103/2231-4040.90879.

Ueda, Eriko, Yutaka Yoshikawa, Yoshio Ishino, Hiromu Sakurai, and Yoshitane Kojima. 2002. "Potential insulinominetic agents of zinc(II) complexes with picolinamide derivatives: Preparations of complexes, in vitro and in vivo studies." *Chemical and Pharmaceutical Bulletin* 50 (3): 337–340. doi:10.1248/cpb.50.337.

Ueda, Minoru, and Shosuke Yamamura. 1999. "The chemistry of leaf-movement in *Mimosa pudica* L." *Tetrahedron* 55 (36): 10937–10948. doi:10.1016/S0040-4020(99)00619-5.

Uhon, Al-Habsi S, and Merell P Billacura. 2018. "Hypoglycemic and protective potentials of the extracts from the air-dried leaves of *Crescentia cujete* Linn." *Science International (Lahore)* 30 (1): 121–125.

Upadhyay, Sukirti, and Prashant Upadhyay. 2011. "*Hibiscus rosa-sinensis*: Pharmacological review." *International Journal of Research in Pharmaceutical and Biomedical Sciences* 2 (4): 1449–1450.

Uppal, Goldie, Vijay Nigam, and Anil Kumar. 2012. "Antidiabetic activity of ethanolic extract of *Euphorbia hirta* Linn." *Der Pharmacia Lettre* 4 (4): 1155–1161.

Uraku, A J, S C Onuoha, C E Offor, M E Ogbanshi, and U S Ndidi. 2011. "The effects of *Abelmoschus Esculentus* fruits on ALP, AST and ALT of diabetic albino rats." *International Journal of Science and Nature* 2 (3): 582 –586.

Useful Tropical Plants. n.d. *Justicia secunda*. Useful Tropical Plants. Accessed October 21, 2020. http://tropical.theferns.info/viewtropical.php?id=Justicia+secunda.

Useful Tropical Plants. n.d. *Tamarindus indica*. Useful Tropical Plants. Accessed August 17, 2020. http://tropical.theferns.info/viewtropical.php?id=Tamarindus+indica.

Usman, Muhammad S, Tariq J Siddiqi, Muhammad M Memon, Muhammad S Khan, Wasiq F Rawasia, Muhammad T Ayub, Jayakumar Sreenivasan, and Yasmeen Golzar. 2018. "Sodium-glucose co-transporter 2 inhibitors and cardiovascular outcomes: A systematic review and meta-analysis." *European Journal of Preventive Cardiology* 25 (5): 495–502. doi:10.1177/2047487318755531.

VA Pharmacy Benefits Management Services; Medical Advisory Panel; VISN Pharmacist Executives. 2015. "Empagliflozin monograph - Empagliflozin (Jardiance)." *National Drug Monograph*, October: 1–16.

Vahid, Hamideh, Hassan Rakhshandeh, and Ahmad Ghorbani. 2017. "Antidiabetic properties of *Capparis spinosa* L. and its components." *Biomedicine & Pharmacotherapy* 92: 293–302. doi:10.1016/j.biopha.2017.05.082.

Vargas-Sánchez, Karina, Edwin Garay-Jaramillo, and Rodrigo E González-Reyes. 2019. "Effects of *Moringa olifiera* on glycaemia and insulin levels: A review of animal and human studies." *Nutrients* 11 (12: 2907): 1–19. doi:10.3390/nu11122907.

Vasudevan, Harish, and John H McNeill. 2007. "Chronic cobalt treatment decreases hyperglycemia in streptozotocin-diabetic rats." *BioMetals* 20 (2): 129–134. doi:10.1007/s10534-006-9020-4.

Venkatachalam, Taracad, V K Kumar, P K Selvi, Avinash O Maske, V Anbarasan, and P S Kumar. 2011. "Antidiabetic activity of *Lantana camara* Linn fruits in normal and streptozotocin-induced diabetic rats." *Journal of Pharmacy Research* 4 (5): 1550–1552.

Venkatesh, Sundararajan, J Thilagavathi, and D Shyam sundar. 2008. "Anti-diabetic activity of flowers of *Hibiscus rosasinensis*." *Fitoterapia* 79 (2): 79–81. doi:10.1016/j.fitote.2007.06.015.

Venkateshwarlu, E, P Dileep, P Rakesh Kumar Reddy, and P Sandhya. 2013. "Evaluation of antidiabetic activity of *Carica papaya* seeds on streptozotocin-induced type-II diabetic rats." *Journal of Advanced Scientific Research* 4 (2): 38–41.

Verma, Sonia, Madhu Gupta, Harvinder Popli, and Geeta Arggawal. 2018. "Diabetes mellitus treatment using herbal drugs." *International Journal of Phytomedicine* 10 (1): 1–10.

Verma, Subodh, and John J V McMurray. 2018. "SGLT2 inhibitors and mechanisms of cardiovascular benefit: A state-of-the-art review." *Diabetologia* 61: 2108–2117.

Vincent, John B. 2003. "The potential value and toxicity of chromium picolinate as a nutritional supplement, weight loss agent and muscle development agent." *Sports Medicine* 33 (3): 213–230. doi:10.2165/00007256-200333030-00004.

Vincent, John B. 2018. "Beneficial effects of chromium(III) and vanadium supplements in diabetes." Chap. 29 in *Nutritional and Therapeutic Interventions for Diabetes and Metabolic Syndrome*, edited by Debasis Bagchi and Nair Sreejayan, 365–374. Tuscaloosa, AL: Academic Press. doi:10.1016/B978-0-12-812019-4.00029-5.

Vinuthan, M K, V G Kumar, J P Ravindra, Jayaprakash, and K Narayana. 2004. "Effect of extracts of *Murraya koenigii* leaves on the levels of blood glucose and plasma insulin in alloxan-induced diabetic rats." *Indian Journal of Physiology and Pharmacology* 48 (3): 348–352.

Virdi, Jaspreet, S Sivakami, S Shahani, A C Suthar, Manisha M Banavalikar, and M K Biyani. 2003. "Antihyperglycemic effects of three extracts from *Momordica charantia*." *Journal of Ethnopharmacology* 88 (1): 107–111. doi:10.1016/s0378-8741(03)00184-3.

Vuksan, Vladimir, John L Sievenpiper, Vernon Y Y Koo, Thomas Francis, Uljana Beljan-Zdravkovic, Zheng Xu, and Edward Vidgen. 2000. "American ginseng (*Panax quinquefolius* L) reduces postprandial glycemia in nondiabetic subjects and subjects with type 2 diabetes mellitus." *Archives of Internal Medicine* 160 (7): 1009–1013. doi:10.1001/archinte.160.7.1009.

Wafer, Rebecca, Panna Tandon, and James E N Minchin. 2017. "The role of peroxisome proliferator-activated receptor gamma (PPARG) in adipogenesis: Applying knowledge from the fish aquaculture industry to biomedical research." *Frontliners in Endocrinology* 8: 1–10. doi:10.3389/fendo.2017.00102.

Waheed, Akbar, Ghulam A Miana, S I Ahmad, and Munir A Khan. 2006. "Clinical investigation of hypoglycemic effect of *Coriandrum Sativum* in type-2 (NIDDM) diabetic patients." *Pakistan Journal of Pharmacology* 23 (1): 7–11.

Wahyuningsih, Doti, and Yudi Purnomo. 2018. "Antidiabetic effect of *Urena lobata*: Preliminary study on hexane, ethanolic, and aqueous leaf extracts." *Jurnal Kedokteran Brawijaya* 30 (1): 1–6.

Walter, Robert M, Janet Y Uriu-Hare, Katherine L Olin, Michelle H Oster, Bradley D Anawalt, James W Critchfield, and Carl L Keen. 1991. "Copper, zinc, manganese, and magnesium status and complications of diabetes mellitus." *Diabetes Care* 14 (11): 1050–1056. doi:10.2337/diacare.14.11.1050.

Wang, George S, and Christopher Hoyte. 2019. "Review of biguanide (metformin) toxicity." *Journal of Intensive Care Medicine* 34 (11–12): 863–876. doi:10.1177/0885066618793385.

Wang, Wei. 2016. "Genomics and traditional Chinese medicine." Chap. 15 in Mats Hansson (Eds.), *Genomics and Society; Ethical, Legal-Cultural, and Socioeconomic Implications*, 293–308. Elsevier Inc.

Wang, Wen, Xu Zhou, Joey S W Kwong, Ling Li, Youping Li, and Xin Sun. 2017. "Efficacy and safety of thiazolidinediones in diabetes patients with renal impairment: A systematic review and meta-analysis." *Scientific Reports* 7 (1): 1–11. doi:10.1038/s41598-017-01965-0.

Wang, Yun-Song, Hong-Ping He, Yue-Mao Shen, Xin Hong, and Xiao-Jiang Hao. 2003. "Two new carbazole alkaloids from *Murraya koenigii*." *Journal of Natural Products* 66 (3): 416–418. doi:10.1021/np020468a.

Wasfi, Rafid M A H, and Yarob S A-J AL-kabi. 2019. "Studying the hypoglycemic activity of celery herb extract *Apium graveolens* in blood glucose level of laboratory rats (Sprague Dawely)." *Journal of Pure and Applied Microbiology* 13 (4): 2389–2395.

Waterman, Carrie, Patricio Rojas-Silva, Tugba B Tumer, Peter Kuhn, Allison J Richard, Shawna Wicks, Jacqueline M Stephens, et al. 2015. "Isothiocyanate-rich *Moringa oleifera* extract reduces weight gain, insulin resistance and hepatic gluconeogenesis in mice." *Molecular Nutrition & Food Research* 59 (6): 1013–1024. doi:10.1002/mnfr.201400679.

Watt, John M, and Maria G Breyer-Brandwijk. 1932. "The medicinal and poisonous plants of Southern Africa." *Edinburgh* 2 (2): 112.

Widharna, R M, A A Soemardji, K R Wiratsutisna, and L B S Kardono. 2010. "Anti Diabetes Mellitus Activity in vivo of Ethanolic Extract and Ethyl Acetate Fraction of *Euphorbia hirta* L. herb." *International Journal of Pharmacology* 6 (3): 231–240.

Wikipedia. 2021. *Curry Tree*. Wikipedia: The Free Encyclopedia. July 5. Accessed August 30, 2021. https://en.wikipedia.org/w/index.php?title=Curry_tree&oldid=1032131478.

Wikipedia. 2021. *Grapefruit*. Wikipedia: The Free Encyclopedia. August 25. Accessed August 29, 2021. https://en.wikipedia.org/w/index.php?title=Grapefruit&oldid=1040639675.

Wikipedia. 2021. *Okra*. Wikipedia: The Free Encyclopedia. August 28. Accessed August 20, 2021. https://en.wikipedia.org/w/index.php?title=Okra&oldid=1041089296.

Wikipedia. 2021. *Phlorizin*. Wikipedia: The Free Encyclopedia. June 6. Accessed July 21, 2021. https://en.wikipedia.org/w/index.php?title=Special:CiteThisPage&page=Phlorizin&id=1027261149&wpFormIdentifier=titleform#:~:text=Permanent%20link%3A-,https%3A//en.wikipedia.org/w/index.php%3Ftitle%3DPhlorizin%26oldid%3D1027261149,-Primary%20contributors%3A.

Wikipedia. 2021. *Saponin*. Wikipedia: The Free Encyclopedia. June 5. https://en.wikipedia.org/wiki/Saponin.

Wikipedia. 2021. *Syzygium cumini*. Wikipedia: The Free Encyclopedia. July 20. Accessed September 4, 2021. https://en.wikipedia.org/w/index.php?title=Syzygium_cumini&oldid=1034553802.

Wikipedia. 2021. *Tungsten*. Wikipedia: The Free Encyclopedia. June 5. https://en.wikipedia.org/wiki/Tungsten.

Wikipedia. 2022. *Sildenafil*. Wikipedia: The Free Encyclopedia. February 22. Accessed March 11, 2022. https://en.wikipedia.org/wiki/Sildenafil.

Willms, Berend, and D Ruge. 1999. "Comparison of acarbose and metformin in patients with type 2 diabetes mellitus insufficiently controlled with diet and sulphonylureas: A randomized, placebo-controlled study." *Diabetic Medicine* 16 (9): 755–761. doi:10.1046/j.1464-5491.1999.00149.x.

Wright, Kirsten M, Janis McFerrin, Armando A Magaña, Janne Roberts, Maya Caruso, Doris Kretzschmar, Jan F Stevens, and Sumyanath, Amala Claudia. 2022. "Developing a rational, optimized product of *Centella asiatica* for examination in clinical trials: Real world challenges." *Frontiers in Nutrition*, 8: 1–18.

Wu, Lei, Wanyue Chen, and Zhang Wang. 2021. "Traditional Indian medicine in China: The status quo of recognition, development and research." *Journal of Ethnopharmacology* 279. doi:10.1016/j.jep.2021.114317.

Wu, Pei-Chen, Vin-Cent Wu, Cheng-Jui Lin, Chi-Feng Pan, Chih-Yang Chen, Tao-Min Huang, Che-Hsiung Wu, Likwang Chen, Chih-Jen Wu, and NRPB Kidney Consortium. 2017. "Meglitinides increase the risk of hypoglycemia in diabetic patients with advanced chronic kidney disease: A nationwide, population-based study." *Oncotarget* 8 (44): 78086–78095.

Xavier, Serene, Jayanarayanan Sadanandan, Naijil George, and Chiramadathikudiyil S Paulose. 2012. "β2-Adrenoceptor and insulin receptor expression in the skeletal muscle of streptozotocin induced diabetic rats: Antagonism by vitamin D3 and curcumin." *European Journal of Pharmacology* 687 (1–3): 14–20. doi:10.1016/j.ejphar.2012.02.050.

Xia, Tao, and Qin Wang. 2007. "Hypoglycaemic role of *Cucurbita ficifolia* (Cucurbitaceae) fruit extract in streptozotocin-induced diabetic rats." *Journal of the Science of Food and Agriculture* 87 (9): 1753–1757. doi:10.1002/jsfa.2916.

Xulu, Sphilile, and Peter M O Owira. 2012. "Naringin ameliorates atherogenic dyslipidemia but not hyperglycemia in rats with type 1 diabetes." *Journal of Cardiovascular Pharmacology* 59 (2): 133–141. doi:10.1097/FJC.0b013e31823827a4.

Yadav, S, V Vats, Y Dhunnoo, and J K Grover. 2002. "Hypoglycemic and antihyperglycemic activity of *Murraya koenigii* leaves in diabetic rats." *Journal of Ethnopharmacology* 82 (2–3): 111–116. doi:10.1016/s0378-8741(02)00167-8.

Yadav, S P, V Vats, A C Ammini, and J K Grover. 2004. "*Brassica juncea* (Rai) significantly prevented the development of insulin resistance in rats fed fructose-enriched diet." *Journal of Ethnopharmacology* 93 (1): 113–116. doi:10.1016/j.jep.2004.03.034.

Yamada, Kazuhiro. 2013. "Cobalt: Its role in health and disease." Vol. 13, chap. 9 in *Interrelations between Essential Metal Ions and Human Diseases: Metal Ions in Life Sciences*, edited by Astrid Sigel, Helmut Sigel and Roland K O Sigel, 295–320. Dordrecht: Springer. doi:10.1007/978-94-007-7500-8_9.

Yang, Luqin, Debbie C Crans, Susie M Miller, Agnete la Cour, Oren P Anderson, Peter M Kaszynski, Michael E Godzala (III), LaTanya D Austin, and Gail R Willsky. 2002. “Cobalt(II) and cobalt(III) dipicolinate complexes: Solid state, solution, and in vivo insulin-like properties.” *Inorganic Chemistry* 41 (19): 4859–4871. doi:10.1021/ic0200621.

Yang, Wen-Chin. 2014. “Botanical, pharmacological, phytochemical, and toxicological aspects of the antidiabetic plant *Bidens pilosa* L.” *Evidence-Based Complementary and Alternative Medicine* 2014: 14. doi:10.1155/2014/698617.

Ybarra, Juan, Alireza Behrooz, Allen Gabriel, Mehmet H Koseoglu, and Faramarz Ismail-Beigi. 1997. “Glycemia-lowering effect of cobalt chloride in the diabetic rat: Increased GLUT1 mRNA expression.” *Molecular and Cellular Endocrinology* 133 (2): 151–160. doi:10.1016/s0303-7207(97)00162-7.

Yokozawa, Takako, Hyun Y Kim, Eun J Cho, Jae S Choi, and Hae Y Chung. 2002. “Antioxidant effects of isorhamnetin 3,7-Di-O-β-D-glucopyranoside isolated from mustard leaf (*Brassica juncea*) in rats with streptozotocin-induced diabetes.” *Journal of Agricultural and Food Chemistry* 50 (19): 5490–5495. doi:10.1021/jf0202133.

Yusni, Yusni, Hendra Zufry, Firdalena Meutia, and Krishna W Sucipto. 2018. “The effects of celery leaf (*Apium graveolens* L.) treatment on blood glucose and insulin levels in elderly pre-diabetics.” *Saudi Medical Journal* 39 (2): 154–160.

Zacharias, N T, K L Sebastian, B Philip, and K T Augusti. 1980. “Hypoglycemic and hypolipidemic effects of garlic in sucrose fed rabbits.” *Indian Journal of Physiology and Pharmacology* 24 (2): 151–154.

Zaidenstein, R, V Dishi, M Gips, S Soback, N Cohen, J Weissgarten, A Blatt, and A Golik. 1998. “The effect of grapefruit juice on the pharmacokinetics of orally administered verapamil.” *European Journal of Clinical Pharmacology* 54 (4): 337–340. doi:10.1007/s002280050470.

Zhang, Chen-Yu, Laura E Parton, Chian P Ye, Stefan Krauss, Ruichao Shen, Cheng-Ting Lin, John A Porco Jr, and Bradford B Lowell. 2006. “Genipin iridoid inhibits UCP2-mediated proton leak and acutely reverses obesity and high glucose-induced beta cell dysfunction in isolated pancreatic islets.” *Cell Metabolism* 3 (6): 417–427. doi:10.1016/j.cmet.2006.04.010.

Zhang, Dong-wei, Min Fu, Si-Hua Gao, and Jun-Li Liu. 2013. “Curcumin and diabetes: A systematic review.” *Evidence-Based Complementary and Alternative Medicine* 2013: 1–16. doi:10.1155/2013/636053.

Zhang, Hongxia, and Zheng F Ma. 2018. “Phytochemical and pharmacological properties of *Capparis spinosa* as a medicinal plant.” *Nutrients* 10 (2): 116–129. doi:10.3390/nu10020116.

Zulcafli, Azrin S, Chooiling Lim, Anna P Ling, Soimoi Chye, and Rhunyian Koh. 2020. “Antidiabetic potential of *Syzygium* sp.: An overview.” *Yale Journal of Biology and Medicine* 93 (2): 307–325.

Abbreviations

A1C	Glycated haemoglobin (HbA1c)
ADP	Adenosine diphosphate
AG	Alpha glucosidase
AGE	Aged garlic extract
AMP	Adenosine monophosphate
AMPK	AMP-activated protein kinase
APDS	Allyl propyl disulphide
ATP	Adenosine triphosphate
BG	Blood glucose
CAT	Catalase activities
CVD	Cardiovascular disease
CYP	Cytochrome P450 enzymes
DIP	Diabetes in pregnancy
DM	Diabetes mellitus
DPP-4	Dipeptidyl peptidase-4
FBG	Fasting blood glucose
FDA	US Food and Drug Administration
FPI	Fasting plasma insulin
GDM	Gestational diabetes mellitus
GI	Gastrointestinal
GIP	Glucose-dependent insulinotropic peptide
GLP-1	Glucagon-like peptide-1
GLUT-4	Glucose transporter type-4
Hb	Haemoglobin
HbA1c	Glycated haemoglobin (A1C)
HDL	High density lipoprotein
HFD	High-fat diet
HG	Hepatic glycogen
HIP	Hyperglycaemia in pregnancy
HMG-CoA	3-hydroxy-3-methyl-glutaryl-coenzyme A
HOMA-IR	Homeostatic model assessment for insulin resistance
HTC	Hepatic total cholesterol
IP	Intraperitoneal
IC50	Half maximal inhibitory concentration
IDF	International Diabetes Federation
INS-R	Insulin substrate receptor
IR-mRNA	Insulin receptor messenger ribonucleic acid

K_{ATP}	ATP-sensitive potassium channel
LD_{50}	The amount of a material, given all at once, which causes the death of 50% (one half) of a group of test animals. The LD_{50} is one way to measure the short-term poisoning potential (acute toxicity) of a material.
LDL	Low density lipoprotein
LPO	Lipid peroxidation
mg	Milligrams
mg/kg.bw	Milligrams per kilogram of body weight
ml/kg.bw	Millilitres per kilogram of body weight
mmol/L	Millimoles per litre
MODY	Maturity onset diabetes of the young
mRNA	Messenger ribonucleic acid
NAC	North American and Caribbean
NIC	Nicotinamide
NIDDM	Non-insulin-dependent diabetes mellitus
OATP1B1	Organic anion transporting polypeptide 1B1 transporter
OGT	Oral glucose tolerance
PBG	Postprandial blood glucose
PDK1	Phosphoinositide-dependent kinase 1
PG	Plasma glucose
PH	Pleckstrin homology
PI3K	Phosphatidylinositol 3-kinase
PIP2	Phosphatidylinositol (4,5)-bisphosphate
PIP3	Phosphatidylinositol (3,4,5)-triphosphate
PKB	Protein kinase-B; Akt
PL	Plasma lipid
po	Per oral
PPAR	Peroxisome proliferator-activated receptor
PPG	Postprandial plasma glucose
ROS	Reactive oxygen species
SGLT	Sodium–glucose cotransporters
SOD	Superoxide dismutase
STZ	Streptozotocin
SUR-1	Sulfonylureas receptor
T1DM	Type-1 diabetes mellitus; type-1 diabetes; type-1 diabetic
T2DM	Type-2 diabetes mellitus; type-2 diabetes; type-2 diabetic
T3 hormone	Triiodothyronine (affects almost every physiological process in the body, including growth and development, metabolism, body temperature, and heart rate)
T4 hormone	Regulates metabolism and plays a crucial role in digestion, muscle function, and bone health
TAIM	Traditional Arabic and Islamic medicine
TC	Total cholesterol
TCM	Traditional Chinese medicine

TG Triglycerides
TIM Traditional Indian medicine
TSH Thyroid-stimulating hormone
TZD Thiazolidinediones
WHO World Health Organization

Glossary

Adipogenesis: The formation of adipocytes (fat cells) from stem cells.

Alkaloids: Any of a class of naturally occurring organic compounds of nitrogen-containing bases.

Allopathic medicine: Pharmaceutical drugs.

Alloxan: An acidic compound obtained by the oxidation of uric acid and isolated as an efflorescent crystalline hydrate. It is known to kill islet cells in the pancreas and induce a diabetic state due to the lack of insulin.

Anthelmintic: Used to destroy parasitic worms.

Antiatherosclerotic: That which counters the effect of atherosclerosis (a disease of the arteries characterized by the deposition of plaques of fatty material on their inner walls).

Antibodies: A protein produced by the immune system to defend the host body from a foreign substance.

Apoptosis: A form of programmed cell death.

Atomic number: Represented by the symbol Z, it is the number of protons in the nucleus of an atom of a particular element.

Ayurvedic medicine: Use of herbs, fruits, and vegetables for medicinal purposes; it may also be extended to include animal, metal, or mineral material.

Curcumin: The main chemical substance of the *Curcuma longa* plant (turmeric). This gives turmeric its bright yellow colour.

Cytokines: Proteins used for cell signalling and communications between cells.

Cytotoxic: Toxic to living cells.

Digitalis: A flowering plant belonging to the family Scrophulariaceae.

Efficacy: The ability to produce a desired or intended result.

Enzymes: Specialized proteins that act as catalysts that help speed up chemical reactions.

Et al. (et alibi): And others.

Ethnomedicine: The study or comparison of traditional medicine based on bioactive compounds in plants and animals and practiced by various ethnic groups.

Etiology: The cause, set of causes, or manner of causation of a disease or condition.

Euglycaemia: A normal blood glucose concentration.

Flavonoids: A group of plant chemicals (phytonutrients) found in the majority of fruit and vegetables.

Gavage: The administration of food or drugs by force, especially to an animal, typically through a tube leading down the throat to the stomach.

Gene: Physical and functional unit of heredity.

Glucose: Monosaccharide (simple sugar) comes from food and provides energy to the body.

Glycaemic index: A number representing the carbohydrates in foods according to how they affect the blood glucose level.

Glycogenesis: Anabolic pathway by which glycogen synthesis occurs from a simpler precursor, glucose 6-phosphate.
Glycoside: A compound formed from a simple sugar and another compound by the replacement of a hydroxyl group in the sugar molecule. Many drugs and poisons derived from plants are glycosides.
Glycosuria: A condition characterized by an excess of sugar in the urine.
HbA1c: Glycated haemoglobin.
Homeostasis: Maintenance of the internal environment of an organism.
Hyperglycaemia: High increase in blood glucose levels.
Hyperlactatemia: Pathological state in which the resting blood lactate concentration is abnormally high.
Hypertrophy: The growth of muscle cells. This allows the size of the muscles to enlarge.
Hypoglycaemia: Decrease in blood glucose levels.
Immunostimulants: Substances (drugs and nutrients) that stimulate the immune system by inducing the activation or increasing the activity of any of its components.
***In vitro*:** Taking place in a test tube, culture dish, or elsewhere outside a living organism.
***In vivo*:** Taking place in a living organism.
Incretin: A metabolic hormone that stimulates a decrease in blood glucose levels.
Insulin: A hormone produced by pancreatic β-cells in the islets of Langerhans that decreases glucose levels or allows a certain amount to pass through the blood.
Insulin-mimetic agents: Agents that have been shown to mimic the actions of insulin.
Insulinoma: A small tumour in the pancreas that produces an excess amount of insulin.
Intraperitoneal: The route of administration of drugs.
Ligand: Ions or molecules that bond to a central metal atom or ion to form a complex.
Lipolysis: The breakdown of fats and other lipids by hydrolysis to release fatty acids.
Macronutrients: Energy-providing chemical substances that come from food and are consumed by organisms in large quantities. They can be subclassified into three groups: carbohydrates, lipids, and proteins.
Mangiferin: An xanthone extracted from different parts of the mango fruit, leaves, stem, and bark.
Micronutrients: All the minerals and vitamins that the body needs.
Modus operandi: An expression referring to "way of working."
Monotherapy: Use of a single drug to treat a particular disorder or disease.
Normoglycaemia: A normal blood glucose concentration.
Normoxic: Having normal oxygen concentrations.
Oleoresin: A natural or artificial mixture of essential oils and a resin, e.g. balsam.
Permeation: Diffusion of molecules through a membrane or the interface of a substance in solution.
Polymorphism: The ability of an object to take on many forms.

Postprandial: During or relating to the period after a meal.

Potency: The strength of an intoxicant or drug, as measured by the amount needed to produce a certain response.

PPAR-γ : A receptor that is of great importance in the regulation of adipogenesis; in addition, it is essential in the prevention of adiposis and the treatment of type-2 diabetes mellitus.

Purgative: A laxative.

Pyruvate: A carboxylic acid produced by the metabolism of glucose.

Streptozotocin:: An agent with specific effects on pancreatic β-cells that is used to induce type-1 diabetes or type-2 diabetes in animal models, such as rats and mice.

Supplement: Something that is added to something to complete or improve it, e.g. nutrients.

Terpenoid: Any of a large class of organic compounds including terpenes, diterpenes, and sesquiterpenes. They have unsaturated molecules composed of linked isoprene units, generally having the formula $(C_5H_8)n$.

Therapeutant: A therapeutic agent.

Tincture: A medicine made by dissolving a drug in alcohol.

Tolbutamide: An oral sulphonylurea hypoglycaemic drug.

Transition element: An element whose atom has a partially filled d subshell.

Type-1 diabetes: Diabetes caused by the low or no production of insulin in the pancreas.

Type-2 diabetes: Diabetes caused by the low production of insulin in the pancreas.

Index

M

N

O

P

R

S

T

V

W

Y

Z

Printed in the United States
by Baker & Taylor Publisher Services

Printed in the United States
by Baker & Taylor Publisher Services